Mathematical Analysis

Mathematical Analysis

Edited by **Victor Nason**

C WILLFORD PRESS

New York

Published by Willford Press,
118-35 Queens Blvd., Suite 400,
Forest Hills, NY 11375, USA
www.willfordpress.com

Mathematical Analysis
Edited by Victor Nason

International Standard Book Number: 978-1-68285-037-4 (Hardback)

The publisher's policy is to use permanent paper from mills that operate a sustainable forestry policy. Furthermore, the publisher ensures that the text paper and cover boards used have met acceptable environmental accreditation standards.

Trademark Notice: Registered trademark of products or corporate names are used only for explanation and identification without intent to infringe.

Printed in the United States of America.

Contents

Preface

Mathematical analysis is one of the most efficient methods of measuring continuous change. A primary application of such analysis models is in economic data. This book unfolds the innovative aspects of mathematical analysis which will be crucial for the progress of this field in the future. Discussed in it are significant concepts of mathematical analysis, such as differentiation, integration, limits, infinite series, etc. along with their econometric applications. The extensive content of this text provides the readers with a thorough understanding of the subject. Students, researchers, experts and all associated with the discipline of mathematics, statistics and economics will benefit alike from this book.

The researches compiled throughout the book are authentic and of high quality, combining several disciplines and from very diverse regions from around the world. Drawing on the contributions of many researchers from diverse countries, the book's objective is to provide the readers with the latest achievements in the area of research. This book will surely be a source of knowledge to all interested and researching the field.

In the end, I would like to express my deep sense of gratitude to all the authors for meeting the set deadlines in completing and submitting their research chapters. I would also like to thank the publisher for the support offered to us throughout the course of the book. Finally, I extend my sincere thanks to my family for being a constant source of inspiration and encouragement.

Editor

1

On the Gap Conjecture concerning group growth

Rostislav Grigorchuk

Abstract   We discuss some new results concerning Gap Conjecture on group growth and present a reduction of it (and its $*$-version) to several special classes of groups. Namely we show that its validity for the classes of simple groups and residually finite groups will imply the Gap Conjecture in full generality. A similar type reduction holds if the Conjecture is valid for residually polycyclic groups and just-infinite groups. The cases of residually solvable groups and right orderable groups are considered as well.

1 Introduction

Growth functions of finitely generated groups were introduced by Schvarz [37] and independently by Milnor [29], and remain popular subject of geometric group theory. Growth of a finitely generated group can be polynomial, exponential or intermediate between polynomial and exponential. The class of groups of polynomial growth coincides with the class of virtually nilpotent groups as was conjectured by Milnor and confirmed by Gromov [24]. Milnor's problem on the existence of groups of intermediate growth was solved by the author in [12, 13], where for any prime p an uncountable family of 2-generated torsion p-groups $\mathcal{G}_\omega^{(p)}$ with different types of intermediate growth was constructed. Here ω is a parameter of construction taking values in the space of infinite sequences over the alphabet on $p + 1$ letters. All groups $\mathcal{G}_\omega^{(p)}$ satisfy the following lower bound on growth function

Communicated by Efim Zelmanov.

The author is partially supported by the Simons Foundation and by NSF grant DMS-1207699.

R. Grigorchuk (✉)
Department of Mathematics, Mailstop 3368, Texas A&M University, College Station, TX 77843-3368, USA
e-mail: grigorch@math.tamu.edu

$$\gamma_{\mathcal{G}_\omega}(n) \succeq e^{\sqrt{n}}, \tag{1.1}$$

where $\gamma_G(n)$ denotes the growth function of a group G and $\succeq$ is a natural comparison of growth functions (see the next section for definition). The inequality (1.1) just indicates that growth of a group is not less than the growth of the function $e^{\sqrt{n}}$.

All groups from families $\mathcal{G}_\omega^{(p)}$ are residually finite-p groups (i.e. are approximated by finite p-groups). In [15] the author proved that the lower bound (1.1) is universal for all residually finite-p groups and this fact has a straightforward generalization to residually nilpotent groups, as it is indicated in [28].

The paper [13] also contains an example of a torsion free group of intermediate growth, which happened to be right orderable group, as was shown in [19]. For this group the lower bound (1.1) also holds.

In the ICM Kyoto paper [23] the author raised a question if the function $e^{\sqrt{n}}$ gives a universal lower bound for all groups of intermediate growth. Moreover, later he conjectured that indeed this is the case. The corresponding conjecture is now called the *Gap Conjecture* on group growth. In this note we collect known facts related to the Conjecture and present some new results. A recent paper [22] gives further information about the history and developments around the notion of growth in group theory.

The first part of the note is introductory. The second part begins with the case of residually solvable groups where basically we present some results of Wilson from [40,42] and a consequence from them. Then we consider the case of right orderable groups, and the final part contains two reductions of the Conjecture (and its ∗-version) to the classes of residually finite groups and simple groups (Theorem 7.4), and to the class of just-infinite groups, modulo its correctness for residually polycyclic groups (Theorem 7.3).

2 Preliminary facts

Let G be a finitely generated group with a system of generators $A = \{a_1, a_2, \ldots, a_m\}$ (throughout the paper we consider only infinite finitely generated groups and only finite systems of generators). The *length* $|g| = |g|_A$ of an element $g \in G$ with respect to A is the length n of the shortest presentation of g in the form

$$g = a_{i_1}^{\pm 1} a_{i_2}^{\pm 1} \cdots a_{i_n}^{\pm 1},$$

where a_{i_j} are elements in A. It depends on the set of generators, but for any two systems of generators A and B there is a constant $C \in \mathbb{N}$ such that the inequalities

$$|g|_A \leq C|g|_B, \qquad |g|_B \leq C|g|_A. \tag{2.1}$$

hold.

The *growth function* of a group G with respect to the generating set A is the function

$$\gamma_G^A(n) = \left|\{g \in G : |g|_A \leq n\}\right|,$$

where $|E|$ denotes the cardinality of a set E, and n is a natural number.

If $\Gamma = \Gamma(G, A)$ is the Cayley graph of a group G with respect to the generating set A, then $|g|$ is the combinatorial distance between vertices g and e (the identity element in G), and $\gamma_G^A(n)$ counts the number of vertices at combinatorial distance $\leq n$ from e (i.e., it counts the number of elements in the ball of radius n with center at the identity element).

It follows from (2.1) that growth functions $\gamma_G^A(n)$, $\gamma_G^B(n)$ satisfy the inequalities

$$\gamma_G^A(n) \leq \gamma_G^B(Cn), \qquad \gamma_G^B(n) \leq \gamma_G^A(Cn). \tag{2.2}$$

The dependence of the growth function on generating set is inconvenience and it is customary to avoid it by using the following trick. Two functions on the naturals γ_1 and γ_2 are called *equivalent* (written $\gamma_1 \sim \gamma_2$) if there is a constant $C \in \mathbb{N}$ such that $\gamma_1(n) \leq C\gamma_2(Cn)$, $\gamma_2(n) \leq C\gamma_1(Cn)$ for all $n \geq 1$. Then according to (2.2), the growth functions constructed with respect to two different systems of generators are equivalent. The class of equivalence $[\gamma_G^A]$ of growth function is called *degree of growth*, or *rate of growth* of G. It is an invariant not only up to isomorphism but also up to weaker equivalence relation called *quasi-isometry* [8].

We will also consider a preoder $\preceq$ on the set of growth functions:

$$\gamma_1(n) \preceq \gamma_2(n) \tag{2.3}$$

if there is an integer $C > 1$ such that $\gamma_1(n) \leq \gamma_2(Cn)$ for all $n \geq 1$. This converts the set $\mathcal{W}$ of growth degrees of finitely generated groups into a partially ordered set. The notation $\prec$ will be used in this article to indicate a strict inequality.

Let us remind some basic facts about growth rates that will be used in the paper.

- The power functions n^α belong to different equivalence classes for different $\alpha \geq 0$.
- The polynomial function $P_d(n) = c_d n^d + \cdots + c_1 n + c_0$, where $c_d \neq 0$ is equivalent to the power function n^d.
- All exponential functions λ^n, $\lambda > 1$ are equivalent and belong to the class $[e^n]$.
- All functions of *intermediate type* e^{n^α}, $0 < \alpha < 1$ belong to different equivalence classes.

This is not a complete list of rates of growth that a group may have. Much more is provided in [12] and [3].

It is easy to see that growth of a group coincides with the growth of a subgroup of finite index, and that growth of a group is not smaller than the growth of a finitely generated subgroup or of a factor group. Since a group with m generators can be presented as a quotient group of a free group of rank m, the growth of a finitely generated group cannot be faster than exponential (i.e., it can not be superexponential). Therefore we can split the growth types into three classes:

- *Polynomial* growth. A group G has *polynomial* growth if there are constants $C > 0$ and $d > 0$ such that $\gamma(n) < Cn^d$ for all $n \geq 1$. Minimal d with this property is called the degree of polynomial growth.
- *Intermediate* growth. A group G has *intermediate* growth if $\gamma(n)$ grows faster than any polynomial but slower than any exponent function λ^n, $\lambda > 1$ (i.e. $\gamma(n) \prec e^n$).
- *Exponential growth.* A group G has *exponential* growth if $\gamma(n)$ is equivalent to e^n.

The question on the existence of groups of intermediate growth was raised in 1968 by Milnor [30]. For many classes of groups (for instance for linear groups by Tits alternative [38], or for solvable groups by the results of Milnor [31] and Wolf [43]) intermediate growth is impossible. Milnor's question was answered by author in 1983 [10,12,20], where it was shown that there are uncountably many 2-generated torsion groups of intermediate growth. Moreover, it was shown in [12,13,20] that for any prime p a partially ordered set $\mathcal{W}_p$ of growth degrees of finitely generated torsion p-groups contains uncountable chain and contains uncountable anti-chain. The immediate consequence of this result is the existence of uncountably many quasi-isometry equivalence classes of finitely generated groups (in fact 2-generated groups) [12].

Below we will use several times the following lemma [24, page 59].

Lemma 2.1 (Splitting lemma) *Let G be a finitely generated group of polynomial growth of degree d and $H \lhd G$ be a normal subgroup with quotient G/H being an infinite cyclic group. Then H has polynomial growth of degree $\leq d - 1$.*

3 Gap Conjecture and its modifications

We will say that a group is *virtually nilpotent* (virtually solvable) if it contains nilpotent (solvable) subgroup of finite index. It was observed around 1968 by Milnor, Wolf, Hartly and Guivarc'h that a nilpotent group has polynomial growth and hence a virtually nilpotent group also has polynomial growth. In his remarkable paper [24], Gromov established the converse.

Theorem 3.1 (Gromov 1981) *If a finitely generated group G has polynomial growth, then G contains a nilpotent subgroup of finite index.*

In fact Gromov obtained stronger result about polynomial growth.

Theorem 3.2 *For any positive integers d and k, there exist positive integers R, N and q with the following property. If a group G with a fixed system of generators satisfies the inequality $\gamma(n) \leq kn^d$ for $n = 1, 2, \ldots, R$ then G contains a nilpotent subgroup H of index at most q and whose degree of nilpotence is at most N.*

The above theorem implies existence of a function υ growing faster than any polynomial and such that if $\gamma_G \prec \upsilon$, then growth of G is polynomial.

Indeed, taking a sequence $\{k_i, d_i\}_{i=1}^{\infty}$ with $k_i \to \infty$ and $d_i \to \infty$ when $i \to \infty$ and the corresponding sequence $\{R_i\}_{i=1}^{\infty}$, whose existence follows from Theorem 3.2, one can build a function $\upsilon(n)$ which coincides with the polynomial $k_i n^{d_i}$ on the interval $[R_{i-1} + 1, R_i]$ and separates polynomial growth from intermediate. Therefore there is a *Gap* in the scale of rates of growth of finitely generated groups and a big problem is to find the optimal function (or at least to provide good lower and upper bounds for it) which separates polynomial growth from intermediate. The best known result in this direction is the function $n^{(\log\log n)^c}$ (c some positive constant) which appeared recently in the paper of Shalom and Tao [36, Corollary 8.6].

The lower bound of the type $e^{\sqrt{n}}$ for all groups $\mathcal{G}_\omega^{(p)}$ of intermediate growth established in [10,12,13,20] allowed the author to guess that equivalence class of function $e^{\sqrt{n}}$ could be a good candidate for a "border" between polynomial and exponential growth. This guess was further strengthened in 1988 when the author obtained the result published in [15] (see Theorem 5.1). For the first time the Gap Conjecture was formulated in the form of a question in 1991 (see [23]).

Conjecture 1 (Gap Conjecture) *If the growth function $\gamma_G(n)$ of a finitely generated group G is strictly bounded from above by $e^{\sqrt{n}}$ (i.e. if $\gamma_G(n) \prec e^{\sqrt{n}}$), then growth of G is polynomial.*

The question of independent interest is whether there is a group, or more generally a cancellative semigroup, with growth equivalent to $e^{\sqrt{n}}$ (for the role of cancellative semigroups in growth business see [14]).

In [22] the author formulated a number of conjectures relevant to the main Conjecture discussed there and in this note. Let us recall some of them as they will play some role in what follow.

Conjecture 2 (Gap Conjecture with parameter β, $0 < \beta < 1$). *If the growth function $\gamma_G(n)$ of a finitely generated group G is strictly bounded from above by e^{n^β} (i.e. if $\gamma(n) \prec e^{n^\beta}$) then the growth of G is polynomial.*

Thus the Gap Conjecture with parameter $1/2$ is just the Gap Conjecture 1. If $\beta < 1/2$ then the Gap Conjecture with parameter β is weaker than the Gap Conjecture, and if $\beta > 1/2$ then it is stronger than the Gap Conjecture.

Conjecture 3 (Weak Gap Conjecture). *There is a β, $0 < \beta < 1$ such that if $\gamma_G(n) \prec e^{n^\beta}$ then the Gap Conjecture with parameter β holds.*

The gap type conjectures can be formulated for other asymptotic characteristics of groups like return probabilities $P_{e,e}^{(n)}$ (e denotes the identity element) for a non degenerate random walk on a group, Følner function $\mathcal{F}(n)$, or spectral density $\mathcal{N}(\lambda)$. There is a close relation between them and the Gap Conjecture on growth, which was mentioned in [22]. When writing this note the author realized that to understand better the relation between different forms of the gap type conjectures it is useful to consider in parallel to the Conjecture 2 [which we will denote $C(\beta)$] a stronger version of it, which we will denote $C^*(\beta)$:

Conjecture 4 (Conjecture $C^*(\beta)$) *If a group C is not virtually nilpotent then $\gamma_C(n) \succeq e^{n^\beta}$.*

It is obvious that $C^*(\beta)$ implies $C(\beta)$ but the opposite is not clear. This is related to the fact that there are groups with incomparable growths [12] as the set $\mathcal{W}$ of rates of growth of finitely generated groups is not linear ordered. The motivation for introducing a $*$-version of the Gap Conjecture will be more clear when a second note [21] of the author is submitted to the arXiv.

4 Growth and elementary amenable groups

Amenable groups were introduced by von Neumann in 1929 [39]. Now they play extremely important role in many branches of mathematics. Let AG denote the class of amenable groups. By a theorem of Adelson-Velskii [1], each finitely generated group of subexponential growth belongs to the class AG. This class contains finite groups and commutative groups and is closed under the following operations:

(1) taking a *subgroup*,
(2) taking a *quotient group*,
(3) *extensions*,
(4) *unions* (i.e. if for some net $\{\alpha\}$, $G_\alpha \in AG$ and $G_\alpha \subset G_\beta$ if $\alpha < \beta$ then $\cup_\alpha G_\alpha \in AG$).

Let EG be the class of *elementary* amenable groups i.e., the smallest class of groups containing finite groups, commutative groups which is closed with respect to the operations (1)–(4). For instance, virtually nilpotent and, more generally, virtually solvable groups belong to the class EG. This concept defined by Day in [6] got further development in the article [5] of Chou who suggested the following approach to study of elementary amenable groups.

For each ordinal α define a subclass EG_α of EG in the following way. EG_0 consists of finite groups and commutative groups. If α is a limit ordinal then

$$EG_\alpha = \bigcup_{\beta \leq \alpha} EG_\beta.$$

Further, $EG_{\alpha+1}$ is defined as as the class of groups which are extensions of groups from set EG_α by groups from the same set or are direct limits of a family of groups from set EG_α. It is known (and easy to check) that each of the classes EG_α is closed with respect to the operations (1) and (2) [5]. By the *elementary complexity* of a group $G \in EG$ we call the smallest α such that $G \in EG_\alpha$.

It was shown in [5] that class EG does not contain groups of intermediate growth, groups of Burnside type (i.e. finitely generated infinite torsion groups), and finitely generated infinite simple groups. A further study of elementary groups and its generalizations was done by Osin [33].

A larger class SG of subexponentially amenable groups was (implicitly) introduced in [9], and explicitly in [16], and studied in [7] and other papers.

A useful fact about groups of intermediate growth which we will use is due to Rosset [35].

Theorem 4.1 *If G is a finitely generated group which does not grow exponentially and H is a normal subgroup such that G/H is solvable, then H is finitely generated.*

We propose the following generalization of this result.

Theorem 4.2 *Let G be a finitely generated group with no free subsemigroup on two generators and let the quotient G/N be an elementary amenable group. Then the kernel N is a finitely generated group.*

The latter two statements and the chain of further statements of the same spirit that appeared in the literature were initiated by the following lemma of Milnor [31]: if G

is a finitely generated group with subexponential growth, and if $x, y \in G$, then the group generated by the set of conjugates $y, xyx^{-1}, x^2yx^{-2}, \ldots$ is finitely generated.

Proof For the proof of the Theorem 4.2 we will apply induction on elementary complexity α of the quotient group $H = G/N$. If complexity is 0 then the group is either finite or abelian. In the first case N is finitely generated for obvious reason. In the second case we apply the following statements from the paper of Longobardi and Rhemtulla [27, Lemmas 1,2].   □

Lemma 4.3 *If G has no free subsemigroups, then for all $a, b \in G$ the subgroup $\langle a^{b^n}, n \in \mathbb{Z} \rangle$ is finitely generated.*

Lemma 4.4 *Let G be a finitely generated group. If $N \trianglelefteq G, G/N$ is cyclic, and $\langle a^{b^n}, n \in \mathbb{Z} \rangle$ is finitely generated for all $a, b \in G$, then N is finitely generated.*

Assume that the statement of the theorem is correct for quotients $H = G/H$ with complexity $\alpha \leq \beta - 1$ for some ordinal $\beta, \beta \geq 1$. The group H, being finitely generated, allows a short exact sequence

$$\{1\} \to A \to H \to B \to \{1\},$$

where $A, B \in EG_{\beta-1}$. Let $\varphi : G \to G/N$ be the canonical homomorphism and $M = \varphi^{-1}(A)$. Then M is a normal subgroup in G and $G/M \simeq G/N/M/N \simeq H/A \simeq B$. By the inductive assumption M is finitely generated and has no free subsemigroup on two generators. As $M/N \simeq A$, again by induction, N is finitely generated and we are done.

We will discuss just-infinite groups in detail in the last section. But let us prove now a preliminary result which will be used later. Recall that a group is called just-infinite if it is infinite, but every proper quotient is finite (i.e. every nontrivial normal subgroup is of finite index). A group G is called hereditary just-infinite if it is residually finite and every subgroup $H < G$ of finite index (including G itself) is just infinite. Observe that a subgroup of finite index of a hereditary just-infinite group is hereditary just-infinite.

We learned the following result from de Cournulier. A proof is provided here as there is no one in the literature.

Theorem 4.5 *Let G be a finitely generated hereditary just-infinite group, and suppose that G belongs to the class EG of elementary amenable groups. Then G is isomorphic either to the infinite cyclic group $\mathbb{Z}$ or to the infinite dihedral group D_∞.*

Proof If $G \in EG_0$ then G is abelian and hence $G \simeq \mathbb{Z}$. Assume that the statement is correct for all groups from classes $EG_\alpha, \alpha < \beta$ for some ordinal β. Let us prove it for β. Assume $G \in EG_\beta$ and β is smallest with this property. β can not be a limit ordinal because G is finitely generated. Therefore G is the extension of a group A by a group $B = G/A$, where $A, B \in EG_{\beta-1}$. In fact B is a finite group (as G is just-infinite). As a subgroup of finite index in a hereditary just-infinite group, A is hereditary just-infinite and moreover finitely generated (as a subgroup of finite index in a finitely generated group). By inductive assumption A is isomorphic either to the infinite cyclic group $\mathbb{Z}$ or to the infinite dihedral group D_∞. In particular G has a normal subgroup H of finite index isomorphic to $\mathbb{Z}$.

Let G act on H by conjugation. Then we get a homomorphism $\psi : G \to Aut(H) \simeq \mathbb{Z}_2$. If $\psi(G) = \{1\}$, then H is a central subgroup. It is a standard fact in group theory (see for instance [25, Proposition 2.4.4]) that if there is a central subgroup of finite index in G then the commutator subgroup G' is finite. But as G is just-infinite, $G' = \{1\}$ and so G is abelian, hence $G \simeq \mathbb{Z}$ in this case.

If $\psi(G) = Aut(H)$ then $N = ker\,\psi$ is a centralizer $C_G(H)$ of H in G. Subgroup N has index 2 in G, is just-infinite and hence by the same reason as above $N' = \{1\}$, so N is abelian. Being finitely generated and just infinite implies $N \simeq \mathbb{Z}$.

Let $x \in G$, $x \notin N$. The element x acts on N by conjugation mapping each element to its inverse. In particular, $x^{-1}(x^2)x = x^{-2}$, so $(x^2)^2 = 1$. But $x^2 \in N$. Since N is torsion free $x^2 = 1$. Therefore

$$G = \langle x, N \rangle = \langle x, y : x^2 = 1, x^{-1}yx = y^{-1} \rangle \simeq D_\infty,$$

where y is a generator of N.   □

5 Gap Conjecture for residually solvable groups

Recall that a group G is said to be a residually finite-p group (sometimes also called residually finite p-group) if it is approximated by finite p-groups, i.e., for any $g \in G$ there is a finite p-group H and a homomorphism $\phi : G \to H$ with $\phi(g) \neq 1$. This class is, of course, smaller than the class of residually finite groups, but it is pretty large. For instance, Golod-Shafarevich groups, p-groups $\mathcal{G}_\omega$ from [12,13], and many other groups belong to this class.

Theorem 5.1 [15] *Let G be a finitely generated residually finite-p group. If $\gamma_G(n) \prec e^{\sqrt{n}}$ then G has polynomial growth.*

As was established by the author in a discussion with Lubotzky and Mann during the conference on profinite groups in Oberwolfach in 1990, the same arguments as given in [12] combined with the following lemma from [28].

Lemma 5.2 (Lemma 1.7, [28]) *Let G be a finitely generated residually nilpotent group. Assume that for every prime p the pro-p-closure $G_{\hat{p}}$ of G is p-adic analytic. Then G is linear.*

allows one to prove a stronger version of the above theorem (see the Remark after Theorem 1.8 in [28]):

Theorem 5.3 *Let G be a residually nilpotent finitely generated group. If $\gamma_G(n) \prec e^{\sqrt{n}}$ then G has polynomial growth.*

To be linear means to be isomorphic to a subgroup of the linear group $GL_n(\mathbb{K})$ for some field $\mathbb{K}$. By Tits alternative [38] every finitely generated linear group either contains a free subgroup on two generators or is virtually solvable. Hence the above lemma immediately reduces Theorem 5.3 to Theorem 5.1.

The latter two theorems (where the first one is the corresponding statement from [15] while the second one is a corrected form of what is stated in Remark on page

527 in [28]) show that Gap Conjecture $C(1/2)$ holds for the class of residually finite-p groups and more generally for the class of residually nilpotent groups. In fact, arguments provided in [15,28] allow to prove stronger conjecture $C^*(1/2)$ for these classes of groups.

Let p be a prime and $a_n^{(p)}$ be the n-th coefficient of the power series given by

$$\sum_{n=0}^{\infty} a_n^{(p)} z^n = \prod_{n=1}^{\infty} \frac{1-z^{pn}}{1-z^n}.$$

Then the lower bound $a_n^{(p)} \succeq e^{\sqrt{n}}$ holds. Moreover if a group G is a residually finite-p group and is not virtually nilpotent then for any system of generators A

$$\gamma_G^A(n) \geq a_n^{(p)}, n = 1, 2, \ldots$$

(see the relation (23) and Lemma 8 in [15]). Observe that the latter statement is valid not only in the case when A is a system of elements that generate G as a group but even in a more general case when A is a generating set for the group G considered as a semigroup. In fact, growth function of any group is bounded from below by a sequence of coefficients of Hilbert-Poincaré series of the universal p-enveloping algebra of the restricted Lie p-algebra associated with the group using the factors of the lower p-central series [15].

Theorem 1.8 from [28] contains an interesting approach to polynomial growth type theorems in the case of residually nilpotent groups. Moreover, as is mentioned in [28] in the remark after the theorem, the proof provided there yields the same conclusion under a weaker assumption: $\gamma_G(n) \prec 2^{2^{\sqrt{\log_2 n}}}$.

Surprisingly, in his first paper on the gap type problem [42] Wilson used a similar upper bound $\gamma_G(n) \prec e^{e^{(1/2)\sqrt{\ln n}}}$ to measure size of a gap for residually solvable groups. Wilson's approach is quite different from those that were used before and is based on exploring self-centralizing chief factors in finite solvable groups.

Recall that a chief factor of a group G is a (nontrivial) minimal normal subgroup of some quotient G/N, and that L/M is a self-centralizing chief factor of a group G if M is normal in G, L/M is a minimal normal subgroup of G/M, and $L/M = C_{G/M}(L/M)$. One of the results in [42] is

Theorem 5.4 (Wilson) *Let G be a residually solvable group of subexponential growth whose finite self-centralizing chief factors all have rank at most k. Then G has a residually nilpotent normal subgroup whose index is finite and bounded in terms of k and $\gamma_G(n)$.*

If, in addition $\gamma_G(n) \prec e^{\sqrt{n}}$, then G has a nilpotent normal subgroup whose index is finite and bounded in terms of k and $\gamma_G(n)$.

The proof of this result is based on the following lemma the proof of which uses ultraproducts.

Lemma 5.5 (Lemma 2.1, [42]) *Let k be a positive integer and $\alpha : \mathbb{N} \to \mathbb{R}_+$ a function such that $\alpha(n)/n \to 0$ as $n \to \infty$. Suppose that G is a finite solvable group having* (i)

a self-centralizing minimal normal subgroup V of rank at most k and (ii) *a generating set A such that* $\gamma_G^A(n) \leq e^{\alpha(n)}$ *for all n. Then* $|G/V|$ *is bounded in terms of k and* α *alone.*

One of the almost immediate corollaries of the technique developed in [42] are the facts stated below in Theorems 5.6 and 5.7.

Recall that a group is called supersolvable if it has a finite normal descending chain of subgroups with cyclic quotients. Every finitely generated nilpotent group is supersolvable [34], and the symmetric group $Sym(4)$ is the simplest example of a solvable but not supersolvable group.

Theorem 5.6 *The Gap Conjecture holds for residually supersolvable groups. Moreover, the conjecture* $C^*(1/2)$ *holds for residually supersolvable groups.*

Developing his technique and using the known facts about maximal primitive solvable subgroups of $GL_n(p)$ (p prime) Wilson in [40] proved that the Gap Conjecture with parameter 1/6 holds for residually solvable groups. In fact what follows from arguments in [42], combined with arguments from [15,28] and with what was written above, can be formulated as

Theorem 5.7 *The conjecture* $C^*(1/6)$ *holds for residually solvable groups.*

There is a hope that eventually the Gap Conjecture and its ∗-version will be proved for residually solvable groups, or at least for residually polycyclic groups (which is the same as to prove it for groups approximated by finite solvable groups, because polycyclic groups are residually finite [34]). If the latter is done, then we will have complete reduction of the Gap Conjecture to just-infinite groups (more on this in the last section).

6 Gap Conjecture for right orderable groups

Recall that a group is called right orderable if there is a linear order on the set of its elements invariant with respect to multiplication on the right. In a similar way are defined left orderable groups. A group is bi-orderable (or totally orderable) if there is a linear order invariant with respect to multiplication on the left and on the right. Every right orderable group is left orderable and vise versa but there are right orderable groups which are not totally orderable (see [26] for examples). As was shown by Machi and the author the class of finitely generated right orderable groups of intermediate growth is nonempty [19]. The corresponding group $\hat{\mathcal{G}}$ was earlier constructed in [16] as an example of a torsion free group of intermediate growth. It was implicitly observed in [19] that the class of countable right orderable groups coincides with the class of groups acting faithfully by homeomorphisms on the line $\mathbb{R}$ (or, what is the same, on the interval [0, 1]). Recently Erschler and Bartholdi managed to compute the growth of $\hat{\mathcal{G}}$ which happens to be $e^{\log(n)n^{\alpha_0}}$ where $\alpha_0 = \log 2/\log(2/\rho) \approx 0.7674$, and ρ is the real root of the polynomial $x^3 + x^2 + x - 2$. The question if there exists a finitely generated, totally orderable group of intermediate growth is still open.

The Gap Conjecture and it modifications stated in Sect. 3 are interesting problems even for the class of right orderable groups. Our next result makes some contribution to this topic. The result of Wilson combined with theorems of Morris [32] and Rosset [35] can be used to prove the following statement.

Theorem 6.1 (i) *The Gap Conjecture with parameter* $1/6$, *and, moreover, the conjecture* $C^*(1/6)$ *hold for right orderable groups.*

(ii) *The Gap Conjecture* $C(1/2)$ [*or its* *-version* $C^*(1/2)$] *holds for right orderable groups if it* [*or its* *-version* $C^*(1/2)$] *holds for residually polycyclic groups.*

Proof (i) Let G be a finitely generated right orderable group with growth $\prec e^{n^{1/6}}$. In [32] Morris proved that every finitely generated right orderable amenable group is indicable (i.e. can be mapped onto $\mathbb{Z}$). As by Adelson-Velskii theorem [1] a group of intermediate growth is amenable, we conclude that the abelianization $G_{ab} = G/[G, G]$ is infinite and hence has a decomposition $G_{ab} = G_{ab}^- \oplus G_{ab}^+$ where $G_{ab}^- \simeq \mathbb{Z}^d, d \geq 1$ is a torsion free part of an abelian group and G_{ab}^+ is a torsion part. Let $N \lhd G$ be a normal subgroup such that $G/N = G_{ab}^-$. Since the commutator subgroup of a group is a characteristic group and the torsion free part of abelian group also is a characteristic subgroup we conclude that N is a characteristic subgroup of G. By Theorem 4.1 N is a finitely generated group. Therefore we can proceed with N as we did with G. This allows us to get a descending chain

$$G > G_1 > G_2 > \cdots \tag{6.1}$$

(where $G_1 = N$ etc) of characteristic subgroups with the property that $G_i/G_{i+1} \simeq \mathbb{Z}^{d_i}$ if $G_{i+1} \neq \{1\}$, for some sequence $d_i \in \mathbb{N}, i = 1, 2, \ldots$.

If the chain (6.1) terminates after finitely many steps then G is solvable and by the results of Milnor and Wolf [31,43] G is virtually nilpotent in this case.

Suppose that chain (6.1) is infinite and consider the intersection $G_\omega = \bigcap_{i=1}^\infty G_i$. If $G_\omega = \{1\}$, then the group G is residually solvable (in fact residually polycyclic), and, because of restriction on growth, by Theorem 5.7, G is virtually nilpotent and hence has polynomial growth of some degree d. But this contradicts Splitting Lemma 2.1. Therefore $G_\omega \neq \{1\}$. G/G_ω is residually polycyclic, has growth not greater than the growth of G and by previous argument is virtually nilpotent. If the degree of polynomial growth of G/G_ω is l then again by Splitting Lemma the length of the chain (6.1) can not be larger than l, and we get a contradiction. The part (i) of the theorem is proven.

Now the proof of part (ii) follows immediately. If we assume that G has growth $\prec e^{\sqrt{n}}$ and that the Gap Conjecture holds for the class of residually polycyclic groups then the arguments from previous part (i) are applicable in the same manner. The only difference is that instead of Theorem 5.7 one should use the assumption that the Gap Conjecture holds for residually polyciclic groups. The same argument works in the case of conjecture $C^*(1/2)$. □

7 Gap Conjecture and just-infinite groups

There is a strong evidence based on considerations presented below that the Gap Conjecture can be reduced to three classes of groups: *simple* groups, *branch* groups and *hereditary just-infinite* groups. These three types of groups appear in a natural partition of the class of just-infinite groups into three subclasses described in Theorem 7.3. The following statement is an easy application of Zorn's lemma.

Proposition 7.1 *Let G be a finitely generated infinite group. Then G has a just-infinite quotient.*

Corollary 7.2 *Let $\mathcal{P}$ be a group theoretical property preserved under taking quotients. If there is a finitely generated group satisfying the property $\mathcal{P}$ then there is a just-infinite group satisfying this property.*

Although the property of a group to have intermediate growth is not preserved when passing to a quotient group (the image may have polynomial growth), by theorems of Gromov [24] and Rosset [35], if the quotient G/H of a group G of intermediate growth has polynomial growth then H is a finitely generated group (of intermediate growth, as the extension of a virtually nilpotent group by a virtually nilpotent group is an elementary amenable group and therefore can not have intermediate growth), and one may look for a just-infinite quotient of H and iterate this process in order to represent G as a consecutive extension of a chain of groups that are virtually nilpotent or just-infinite groups. This observation was used in the previous section for the proof of Theorem 6.1 and is the base of the arguments for Theorems 7.4 and 7.5.

Recall that hereditary just-infinite groups were already defined in Sect. 4. We call a just infinite group near simple if it contains a subgroup of finite index which is a direct product of finitely many copies of a simple group.

Branch groups are groups that have a faithful level transitive action on an infinite spherically homogeneous rooted tree $T_{\bar{m}}$ defined by a sequence $\{m_n\}_{n=1}^{\infty}$ of natural numbers $m_n \geq 2$ (determining the branching number for vertices of level n) with the property that the rigid stabilizer $rist_G(n)$ has finite index in G for each $n \geq 1$. Here by $rist_G(n)$ we mean a subgroup $\prod_{v \in V_n} rist_G(v_n)$ which is a product of rigid stabilizers $rist_G(v_n)$ of vertices v_n taken over the set V_n of all vertices of level n, and $rist_G(v)$ is a subgroup of G consisting of elements fixing the vertex v and acting trivially outside the full subtree with the root at v. For a more detailed discussion of this notion see [4, 18]. This is a geometric definition. It follows immediately from the definition that branch groups are infinite. The definition of an algebraically branch group can be found in [4, 17]. Every geometrically branch group is algebraically branch but not vice versa. If G is algebraically branch then it has a quotient G/N which is geometrically branch. The difference between two versions of the definitions is not large but still there is no complete understanding how much the two classes differ (it is not clear what can be said about the kernel N, it is believed that it should be central in G). For just-infinite branch groups the algebraic and geometric definitions are equivalent. Not every branch group is just-infinite, but every proper quotient of a branch group is virtually abelian [18]. Therefore branch groups are "almost just-infinite" and most of known finitely generated branch groups are just-infinite. Observe that a finitely generated virtually nilpotent group is not branch. This follows for instance from the fact that a finitely generated nilpotent group satisfies a minimal condition for normal subgroups while a branch group not.

The next theorem was derived by the author from a result of Wilson [41].

Theorem 7.3 [18] *The class of just-infinite groups naturally splits into three subclasses: (B) branch just-infinite groups, (H) hereditary just-infinite groups, and (S) near-simple just-infinite groups.*

It is already known that there are finitely generated branch groups of intermediate growth. For instance, groups $\mathcal{G}_\omega$ of intermediate growth from the articles [11,13] are of this type. In fact, all known examples of groups of intermediate growth are of branch type or are reconstructions on the base of groups of branch type. The question about existence of amenable but non-elementary amenable hereditary just-infinite group is still open (remind that by Theorem 4.5 the only elementary amenable hereditary just-infinite groups are $\mathbb{Z}$ and D_∞).

Problem 1 Are there finitely generated hereditary just-infinite groups of intermediate growth?

Problem 2 Are there finitely generated simple groups of intermediate growth?

The next theorem is a straightforward corollary of the main result of Bajorska and Makedonska from [2] (observe that it was not stated in [2]). Here we suggest a different proof which is adapted to the needs of the proof of the main Theorem 7.5.

Theorem 7.4 *If the Gap Conjecture or conjecture $C^*(1/2)$ holds for the classes of residually finite groups and simple groups, then the corresponding conjecture holds for the class of all groups.*

Proof Assume that the Gap Conjecture is correct for residually finite groups and for simple groups. Let G be a finitely generated group with growth $\prec e^{\sqrt{n}}$. By Proposition 7.1 it has just-infinite quotient $\bar{G} = G/N$, which belongs to one of the three types of groups listed in the statement of the Theorem 7.3. The rate of growth of $\bar{G}$ is not greater than the rate of growth of $e^{\sqrt{n}}$. The group $\bar{G}$ can not be near simple because in this case it will have a subgroup H of finite index with infinite finitely generated simple quotient whose rate of growth is $\prec e^{\sqrt{n}}$. This is impossible as a virtually nilpotent group can not be infinite simple.

The group $\bar{G}$ also can not be branch as branch groups are residually finite and finitely generated virtually nilpotent groups are not branch. So we can assume that $\bar{G}$ is hereditary just infinite and hence residually finite. Using the assumption of the theorem we conclude that $\bar{G}$ is virtually nilpotent, and therefore elementary amenable. By Theorem 4.5 $\bar{G}$ is isomorphic either to the infinite cyclic group or to the infinite dihedral group D_∞. By Theorem 4.1 kernel N is finitely generated. As the rate of growth of N is less than $e^{\sqrt{n}}$ we can apply to N the same arguments as for G in order to get a surjective homomorphism either onto $\mathbb{Z}$ or onto D_∞.

If $G/N \simeq \mathbb{Z}$, then we repeat the first step of the proof of Theorem 6.1 replacing N by a finitely generated characteristic subgroup $N_1 \lhd G$ with quotient $G/N_1 \simeq \mathbb{Z}^{d_1}$ for some $d_1 \geq 1$. If $G/N_1 \simeq D_\infty$ then we slightly modify the first step. Namely, in this case G has indicable subgroup H of index 2. Let H_1 be the intersection of groups $H^\phi, \phi \in Aut(G)$. As there are only finitely many subgroups of index 2 in G this intersection involves only finitely many groups and H_1 is a characteristic subgroup in

G of finite index of type 2^t for some $t \in \mathbb{N}$. Moreover, $G/H_1 \simeq \mathbb{Z}_2^t$ as the quotient G/H_1 is isomorphic to a subgroup of a direct product of finitely many copies of group $\mathbb{Z}_2$ of order 2. The subgroup H_1, being a subgroup of index 2^{t-1} in H, is indicable and we can apply the argument of the first step of the proof of Theorem 6.1 getting a finitely generated subgroup $H_2 \trianglelefteq H_1$ characteristic in G with quotient $H_1/H_2 \simeq \mathbb{Z}^{d_1}$ for some $d_1 \in \mathbb{N}$.

Let $G_1 \lhd G$ be a subgroup N, H_1 or H_2 depending on the case. Proceed with G_1 in a similar fashion as we did with G, etc. We get a descending chain $\{G_i\}_{i \geq 1}$ of finitely generated subgroups characteristic in G. There are two possibilities.

(1) After finitely many steps we get a group G_i which is hereditary just-infinite and elementary amenable, and hence infinite cyclic or D_∞ (Theorem 4.5). In this case G is polycyclic and we are done in view of the result of Milnor and Wolf on growth of solvable groups.

(2) The process of construction of the chain of subgroups will continue forever. In this case we get a chain with the property that G_i/G_{i+1} is isomorphic either to
    (i) $\mathbb{Z}^{d_i}$, $d_i \in \mathbb{N}$ or to
    (ii) $\mathbb{Z}_2^{t_i}$, $t_i \in \mathbb{N}$. Moreover, each step of type (ii) is immediately followed by a step of type (i).

Let us show that this is impossible. Let G_ω be the intersection $\bigcap_{i \geq 1} G_i$. Then G/G_ω is residually polycyclic and hence residually finite as every polycyclic group is residually finite [34]. Growth of G/G_ω is less than $e^{\sqrt{n}}$. Hence by the assumption of the theorem the group G/G_ω is virtually nilpotent with the rate of polynomial growth of degree d for some $d \in \mathbb{N}$. But this contradicts the splitting lemma as for infinitely many i the quotients G_i/G_{i+1} are isomorphic to $\mathbb{Z}^{d_i}$. This proves the conjecture $C(1/2)$.

In the case of the conjecture $C^*(1/2)$ we proceed in a similar fashion. Only at the beginning we assume that the conjecture $C^*(1/2)$ holds for residually finite groups and for simple groups and that G is a finitely generated group of intermediate growth whose growth does not satisfy inequality $\gamma(n) \succeq e^{n^{1/2}}$. $\qquad\square$

Now we state and prove our main result.

Theorem 7.5 (i) *If the Gap Conjecture with parameter $1/6$ or its $*$-version $C^*(1/6)$ holds for just-infinite groups then the corresponding conjecture holds for all groups.*

(ii) *If the Gap Conjecture or its $*$-version $C^*(1/2)$ holds for residually polycyclic groups and for just-infinite groups then the corresponding conjecture holds for all groups.*

Proof (i) The proof follows the same strategy as the proof of Theorem 7.4. Let G be a finitely generated group with growth $\prec e^{n^{1/6}}$. There can be two possibilities.

(1) G has a finite descending chain $\{G_i\}_{i=1}^k$ of finitely generated characteristic in G groups with consecutive quotients $G_i/G_{i+1} \simeq \mathbb{Z}^{d_i}$ or $G_i/G_{i+1} \simeq \mathbb{Z}_2^{t_i}$, for $i < k$ and $G_k = \{1\}$. In this case G is polycyclic and hence virtually nilpotent

(2) G has an infinite descending chain $\{G_i\}_{i=1}^{\infty}$, with the property that $G_i/G_{i+1} \simeq \mathbb{Z}^{d_i}$ or $G_i/G_{i+1} \simeq \mathbb{Z}_2^{t_i}$, and if $G_i/G_{i+1} \simeq \mathbb{Z}_2^{t_i}$ then $G_{i+1}/G_{i+2} \simeq \mathbb{Z}^{d_{i+1}}$. The group G/G_{ω}, where $G_{\omega} = \bigcap_{i \geq 1} G_i$, is residually polycyclic with growth $\prec e^{n^{1/6}}$. Apply in this case the result of Wilson stated in Theorem 5.4 concluding that G/G_{ω} is virtually nilpotent which is impossible by the splitting lemma.

(ii) Proceed as in (i) with the only difference that in the subcase (2) we apply the assumption that the Gap Conjecture holds for residually polycyclic groups to conclude that this subcase is impossible.

These are arguments for $C(1/2)$ version. The arguments for $*$-version $C^*(1/2)$ are similar. $\square$

Acknowledgments   This work was completed during visit of the author to the Institute Mittag-Leffler (Djursholm, Sweden) associated with the program "Geometric and Analytic Aspects of Group Theory". The author acknowledges organizers of this program. Also the author would like to thank A. Mann for indication of the article [2], and I. Bondarenko and E. Zelmanov for numerous valuable remarks concerning the first draft of this note.

References

1. Adelson-Velskiĭ, G.M., Šreĭder, Y.A.: The Banach mean on groups. Uspehi Mat. Nauk (N.S.), 12(6), 131–136 (1957)
2. Bajorska, B., Macedońska, O.: A note on groups of intermediate growth. Commun. Algebra **35**(12), 4112–4115 (2007)
3. Bartholdi, L., Erschler, A.: Groups of given intermediate word growth, 2011
4. Bartholdi, L., Grigorchuk, R.I., Šunik, Z.: Branch groups. In: Handbook of algebra, vol. 3, pp. 989–1112. North-Holland, Amsterdam (2003)
5. Ching, C.: Elementary amenable groups. Ill. J. Math. **24**(3), 396–407 (1980)
6. Day, M.M.: Amenable semigroups. Ill. J. Math. **1**, 509–544 (1957)
7. de la Harpe, P., Grigorchuk, R.I., Ceccherini-Silberstein, T.: Amenability and paradoxical decompositions for pseudogroups and discrete metric spaces. Tr. Mat. Inst. Steklova 224(Algebra. Topol. Differ. Uravn. i ikh Prilozh.), 68–111 (1999)
8. de la Harpe, P.: Topics in geometric group theory. Chicago Lectures in Mathematics. University of Chicago Press, Chicago (2000)
9. Freedman, M.H., Teichner, P.: 4-manifold topology. I. Subexponential groups. Invent. Math. **122**(3), 509–529 (1995)
10. Grigorchuk, R.I.: On the Milnor problem of group growth. Dokl. Akad. Nauk SSSR **271**(1), 30–33 (1983)
11. Grigorchuk, R.I.: Construction of p-groups of intermediate growth that have a continuum of factor-groups. Algebra i Logika **23**(4), 383–394, 478 (1984)
12. Grigorchuk, R.I.: Degrees of growth of finitely generated groups and the theory of invariant means. Izv. Akad. Nauk SSSR Ser. Mat. **48**(5), 939–985 (1984)
13. Grigorchuk, R.I.: Degrees of growth of p-groups and torsion-free groups. Mat. Sb. (N.S.), **126**(2), 194–214, 286 (1985)
14. Grigorchuk, R.I.: Semigroups with cancellations of polynomial growth. Mat. Zametki **43**(3), 305–319, 428 (1988)
15. Grigorchuk, R.I.: On the Hilbert-Poincare series of graded algebras that are associated with groups. Mat. Sb. **180**(2), 207–225, 304 (1989)

16. Grigorchuk, R.I.: An example of a finitely presented amenable group that does not belong to the class EG. Mat. Sb. **189**(1), 79–100 (1998)
17. Grigorchuk, R.I.: Branch groups. Mat. Zametki **67**(6), 852–858 (2000)
18. Grigorchuk, R.I.: Just infinite branch groups. In: New horizons in pro-p groups, volume 184 of Progr. Math., pp. 121–179. Birkhauser Boston, Boston (2000)
19. Grigorchuk, R.I., Machi, A.: On a group of intermediate growth that acts on a line by homeomorphisms. Mat. Zametki **53**(2), 46–63 (1993)
20. Grigorchuk, R.I.: Groups with intermediate growth function and their applications. Habilitation, Steklov Institute of Mathematics (1985)
21. Grigorchuk, R.: The gap type conjectures for various asymptotic characteristics of groups (in preparation)
22. Grigorchuk, R.: Milnor's problem on the growth of groups and its consequences (2011)

23. Grigorchuk, R.I.: On growth in group theory. In Proceedings of the International Congress of Mathematicians, vols. I, II (Kyoto, 1990), pp. 325–338. Math. Soc. Japan, Tokyo (1991)
24. Mikhael, G.: Groups of polynomial growth and expanding maps. Inst. Hautes Études Sci. Publ. Math. **53**, 53–73 (1981)
25. Gregory, K.: The Schur multiplier, volume 2 of London Mathematical Society Monographs. New Series. The Clarendon Press Oxford University Press, New York (1987)
26. Ivanovič Kokorin, A., Matveevič Kopytov, V.: Fully ordered groups. Halsted Press/Wiley, New York/Toronto (1974) (Translated from the Russian by D. Louvish)
27. Longobardi, P., Maj, M., Rhemtulla, A.H.: Groups with no free subsemigroups. Trans. Am. Math. Soc. **347**(4), 1419–1427 (1995)
28. Lubotzky, A., Mann, A.: On groups of polynomial subgroup growth. Invent. Math. **104**(3), 521–533 (1991)
29. Milnor, J.: A note on curvature and fundamental group. J. Differ. Geom. **2**, 1–7 (1968)
30. Milnor, J.: Problem 5603. Am. Math. Monthly **75**, 685–686 (1968)
31. Milnor, J.: Growth of finitely generated solvable groups. J. Differ. Geom. **2**, 447–449 (1968)
32. Morris, D.W.: Amenable groups that act on the line. Algebr. Geom. Topol. **6**, 2509–2518 (2006)
33. Osin, D.V.: Algebraic entropy of elementary amenable groups. Geom. Dedicata **107**, 133–151 (2004)
34. Robinson, D.J.S.: A course in the theory of groups, 2nd edn, volume 80 of Graduate Texts in Mathematics. Springer, New York (1996)
35. Rosset, S.: A property of groups of non-exponential growth. Proc. Am. Math. Soc. **54**, 24–26 (1976)
36. Shalom, Y., Tao, T.: A finitary version of Gromov's polynomial growth theorem. Geom. Funct. Anal. **20**(6), 1502–1547 (2010)
37. Švarc, A.: A volume invariant of covering. Dokl. Akad. Nauj SSSR **105**, 32–34 (1955)
38. Tits, J.: Free subgroups in linear groups. J. Algebra **20**, 250–270 (1972)
39. von Neumann, J.: Zurr allgemeinen theorie des masses. Fundam. Math. **13**, 73–116 (1929)
40. Wilson, J.: The gap in the growth of residually soluble groups. Bull. London Math. Soc. **2**, 12
41. Wilson, J.S.: Groups with every proper quotient finite. Proc. Camb. Philos. Soc. **69**, 373–391 (1971)
42. Wilson, J.S.: On the growth of residually soluble groups. J. Lond. Math. Soc. (2) **71**(1), 121–132 (2005)
43. Wolf, J.A.: Growth of finitely generated solvable groups and curvature of Riemanniann manifolds. J. Differ. Geom. **2**, 421–446 (1968)

2

Topological stability of a sequence of maps on a compact metric space

Dhaval Thakkar · Ruchi Das

Abstract In this paper we discuss the dynamical system induced by sequence of maps i.e. time varying map on a metric space. We define and study shadowing and expansiveness of such dynamical systems. We show that expansiveness and shadowing of time varying maps are conjugacy invariant. Finally, we prove that a time varying map having shadowing and expansiveness is topologically stable in the class of all time varying maps on a compact metric space.

Keywords Expansiveness · Shadowing · Conjugacy · Topological stability

Mathematics Subject Classification (2010) Primary 54H20; Secondary 37C75 · 37C15

1 Introduction

Expansiveness and shadowing are very important and useful dynamical properties of maps on metric spaces. They have lots of applications in Topological dynamics, Ergodic theory, Symbolic dynamics and related areas. One can refer [2,20] for detailed study of these notions. The concept of expansiveness originally introduced for homeomorphisms on metric spaces [24] has been generalized to positive expansiveness [9], point-wise expansiveness [21], entropy-expansiveness [3], continuum-wise expansiveness [13], measure expansiveness [17] and n-expansiveness [15]. Various

Communicated by S.K. Jain.

D. Thakkar · R. Das (✉)
Department of Mathematics, Faculty of Science,
The M. S. University of Baroda, Vadodara 390002, India
e-mail: rdasmsu@gmail.com

kinds of shadowing also have been defined and their equivalences have been studied in [20,14].

In [25], Walters has introduced the concept of topological stability and proved that Anosov diffeomorphisms are topologically stable. Expansiveness and shadowing play important role in the study of topological stability of maps on a compact metric space [26]. In [19], Nitecki has shown that topological stability is a necessary condition to get axiom A together with strong transversality. Morse-Smale flows are topologically stable is proved by Robinson in [22]. In [10–12], Hurley has obtained necessary conditions for topological stability. Moriyasu [16] has proved that the C^1-interior of the set of all topologically stable diffeomorphisms is characterized as the set of all C^1-structurally stable diffeomorphisms. In [18], authors have proved that, if X^t is a flow in the C^1-interior of the set of topologically stable flows, then X^t satisfies the Axiom A and the strong transversality condition. In [5], authors have proved similar result for the class of incompressible flows and also for volume-preserving diffeomorphisms. In [4,5], authors have generalized results of [5,18] for symplectomorphisms. Recently in [8], authors have studied expansiveness, shadowing, topological stability and decomposition theorems for homeomorphisms on non-compact and non-metrizable spaces.

In discrete dynamical system (X, f), where X is a metric space and $f : X \to X$ is a continuous map, we consider the iterates of points of X under the action of f with discrete ticks of time. Here we consider the case when the function f is changing with the ticks of time. i.e. we consider the action of sequence of functions $\{f_n\}_{n=0}^{\infty}$, where we always consider f_0 to be the identity map. We denote this action by F and call (X, F) a time varying dynamical system. For example, any moving picture on a television screen is an example of time varying dynamical system. In fact television screen is divided into pixels each of a single color red,blue or green. Also if (X, σ) is a shift-space and $\{t_n\}$ is a sequence of integers then $\{\sigma^{t_n}\}$ is a time varying map on X. Similar kind of study related to random perturbations of dynamical systems has been done by Araújo [1]. In [23], introducing many new concepts authors have defined and studied chaos of a time varying map. i.e. of a time varying dynamical system. In [7], author has studied G-chaos of a sequence of maps in a metric G-space.

In this paper, we extend the notion of expansiveness and shadowing for time varying maps on a metric space and study them in detail. In Sect. 2, we define and study expansiveness of a time-varying map on a metric space. In Sect. 3, we define and study shadowing or P.O.T.P. for a time varying map on a metric space. In Sect. 4, we study topological stability of a time varying map on a compact metric space.

2 Expansiveness of a time varying map

Throughout we consider (X, d) to be a metric space and $f_n : X \to X$ to be a sequence of continuous maps, $n = 0, 1, 2, \ldots$ and $F = \{f_n\}_{n=0}^{\infty}$ be a time varying map on X. We denote

$$F_n = f_n \circ f_{n-1} \circ \cdots \circ f_1 \circ f_0, \quad \text{for all} \quad n = 0, 1, 2 \ldots$$

For any $i \leq j$ we define

$$F_{[i,j]} = f_j \circ f_{j-1} \circ \cdots \circ f_{i+1} \circ f_i$$

and for $i > j$, we define $F_{[i,j]}$ to be the identity map on X. For any $k > 0$, we define a time varying map (k^{th}-iterate of F) $F^k = \{g_n\}_{n=0}^\infty$ on X, where

$$g_n = f_{nk} \circ f_{(n-1)k+k-1} \circ \cdots \circ f_{(n-1)k+2} \circ f_{(n-1)k+1} \quad for \ all \ \ n \geq 0.$$

Thus $F^k = \{F_{[(n-1)k+1,nk]}\}_{n=0}^\infty$.

Definition 2.1 [23] Let (X, d) be a metric space and $f_n : X \to X$ be a sequence of maps, $n = 0, 1, 2, \ldots$. For a point $x_0 \in X$, define a sequence as follows:

$$x_{n+1} = f_{n+1}(x_n), \quad n = 0, 1, 2, \ldots$$

Then the sequence $O(x_0) = \{x_n\}_{n=0}^\infty$ is said to be the orbit of x_0 under time varying map (or sequence of maps) $F = \{f_n\}_{n=0}^\infty$.

Definition 2.2 Let (X, d) be a metric space and $f_n : X \to X$ be a sequence of maps, $n = 0, 1, 2, \ldots$. The time varying map $F = \{f_n\}_{n=0}^\infty$ is said to be expansive if there exists a constant $c > 0$ (called an expansive constant) such that for any $x, y \in X$, $x \neq y, d(F_n(x), F_n(y)) > c$ for some $n \geq 0$. Equivalently, if for $x, y \in X$, $d(F_n(x), F_n(y)) \leq c$ for all $n \geq 0$ then $x = y$.

Remark 2.1 If in the above definition $f_n = f$ for all $n \geq 0$, where $f : X \to X$ is continuous, then expansiveness of time varying map $F = \{f_n\}_{n=0}^\infty$ on X is equivalent to positive-expansiveness of f on X [9].

Remark 2.2 Note that expansiveness of a time variant map F is independent of choice of metric if X is compact. Let d_1 and d_2 be two equivalent metrics on a compact space X. Suppose F is expansive on (X, d_1) with expansive constant $\varepsilon > 0$. Since d_1 is equivalent to d_2, there exists an $\varepsilon_1 > 0$ such that for any $x \in X$, $N_{d_2}(x, \varepsilon_1) \subset N_{d_1}(x, \varepsilon)$, where $N_{d_i}(z, \delta)$ denotes open ball centred at z in X of radius δ under metric d_i, $i = 1, 2$. Since X is compact, ε_1 depends only on ε not on x. Let $x \neq y$. Since F is expansive in (X, d_1) with expansive constant $\varepsilon > 0$, $F_n(y) \notin N_{d_1}(F_n(x), \varepsilon)$ for some $n \geq 0$. Now since $N_{d_2}(F_n(x), \varepsilon_1) \subset N_{d_1}(F_n(x), \varepsilon)$, $F_n(y) \notin N_{d_2}(F_n(x), \varepsilon_1)$. Thus F is expansive on (X, d_2) with expansive constant ε_1.

Example 2.1 Consider the time varying map $F = \{f_n\}_{n=0}^\infty$ on the real line $\mathbb{R}$ defined by $f_n(x) = (n + 1)x$, for $x \in \mathbb{R}$ and $n \geq 0$.

Choose $c > 0$. Then for $x, y \in \mathbb{R}$, $x \neq y$, there exists $n \geq 0$ such that $|F_n(x) - F_n(y)| = (n + 1)!|x - y| > c$. Thus F is expansive with expansive constant c.

Definition 2.3 If $h : X \to Y$ is a homeomorphism , h is uniformly continuous on X and h^{-1} is uniformly continuous on Y, then h is said to be a uniform homeomorphism.

Definition 2.4 Let (X, d_1) and (Y, d_2) be two metric spaces. Let $F = \{f_n\}_{n=0}^{\infty}$ and $G = \{g_n\}_{n=0}^{\infty}$ be time varying maps on X and Y respectively. If there is a homeomorphism $h : X \to Y$ such that $h \circ f_n = g_n \circ h$ for all $n = 0, 1, 2, \ldots$ then F and G are said to be conjugate (with respect to the map h) or h-conjugate. In particular, if $h : X \to Y$ is a uniform homeomorphism then F and G are said to be uniformly conjugate or uniformly h-conjugate.

For example, if $F = \{x^{n+1}\}_{n=0}^{\infty}$ on $[0, 1]$, $G = \{2((x+1)/2)^{n+1} - 1\}_{n=0}^{\infty}$ on $[-1, 1]$ then F is uniformly h-conjugate to G, where $h : [0, 1] \to [-1, 1]$ is defined by $h(x) = 2x - 1$.

Theorem 2.1 *Let (X, d_1) and (Y, d_2) be metric spaces. Let $F = \{f_n\}_{n=0}^{\infty}$ and $G = \{g_n\}_{n=0}^{\infty}$ be time varying maps on X and Y respectively such that F is uniformly conjugate to G. If F is expansive on X then G is expansive on Y.*

Proof Let $\varepsilon > 0$ be an expansive constant for F. Since F is uniformly conjugate to G, there exists a uniform homeomorphism $h : X \to Y$ such that $h \circ f_n = g_n \circ h$ for all $n \geq 0$. i.e. $f_n \circ h^{-1} = h^{-1} \circ g_n$ for all $n \geq 0$ which implies $F_n \circ h^{-1} = h^{-1} \circ G_n$ for all $n \geq 0$. Now, h being a uniform homeomorphism, h^{-1} is uniformly continuous therefore for $\varepsilon > 0$ there exists a $\delta > 0$ such that for $x, y \in Y$, $d_2(x, y) < \delta$ implies $d_1(h^{-1}(x), h^{-1}(y)) < \varepsilon$. Let $x, y \in Y$. If for all $n \geq 0$ $d_2(G_n(x), G_n(y)) < \delta$ then $d_1(h^{-1}(G_n(x)), h^{-1}(G_n(y))) < \varepsilon$ for all $n \geq 0$. i.e. $d_1(F_n(h^{-1}(x)), F_n(h^{-1}(y))) < \varepsilon$ for all $n \geq 0$. Since F is expansive with expansive constant ε, we get $h^{-1}(x) = h^{-1}(y)$ which implies $x = y$. Thus G is expansive with expansive constant δ.

Corollary 2.1 *Let (X, d_1) be a compact metric space, (Y, d_2) be a metric space, $F = \{f_n\}_{n=0}^{\infty}$ be a time varying map on X and $h : X \to Y$ is a homeomorphism. If F is expansive on X then $G = h \circ F \circ h^{-1} = \{h \circ f_n \circ h^{-1}\}_{n=0}^{\infty}$ is expansive on Y.*

Theorem 2.2 *Let (X, d) be a compact metric space, $\{f_n\}_{n=0}^{\infty}$ be an equicontinuous family of self maps on X and k be a positive integer. Then time varying map $F = \{f_n\}_{n=0}^{\infty}$ is expansive if and only if F^k is expansive.*

Proof Let $e > 0$ be an expansive constant for F. Since $\{f_n\}_{n=0}^{\infty}$ is equicontinuous family, for any $n > 0$ and $nk + 1 \leq j \leq (n+1)k$, $F_{[nk+1,j]}$ is uniformly continuous on X and therefore there exists a $\delta_j > 0$ such that

$$d(x, y) < \delta_j \Rightarrow d(F_{[nk+1,j]}(x), F_{[nk+1,j]}(y)) < e.$$

Note that due to equicontinuity of $\{f_n\}_{n=0}^{\infty}$, δ_j does not depend on n. Take $\delta = \min\{\delta_j : nk + 1 \leq j \leq (n+1)k\}$. Then for any $n \geq 0$,

$$d(x, y) < \delta \Rightarrow d(F_{[nk+1,j]}(x), F_{[nk+1,j]}(y)) < e.$$

Now $F^k = \{g_n\}_{n=0}^{\infty}$, where $g_n = F_{[(n-1)k+1,nk]}$ and $G_n = g_n \circ \cdots \circ g_1 \circ g_0$. It is easy to see that $G_n = F_{nk}$. Note that for any $j \geq 0$ there exists $n \geq 0$ such that $nk \leq j \leq (n+1)k$. Now for any $n \geq 0$ and $nk \leq j \leq (n+1)k$,

$$d(G_n(x), G_n(y)) < \delta \Rightarrow d(F_{nk}(x), F_{nk}(y)) < \delta$$
$$\Rightarrow d(F_{[nk+1,j]}(F_{nk}(x)), F_{[nk+1,j]}(F_{nk}(y))) < e$$
$$\Rightarrow d(F_j(x), F_j(y)) < e.$$

Since e is an expansive constant for F, $x = y$ and hence δ is an expansive constant for F^k.

Conversely, if F^k is expansive with an expansive constant ε then for any $x, y \in X$, $x \neq y$, there exists $n \geq 0$ such that $d(G_n(x), G_n(y)) > \varepsilon$ which implies $d(F_{nk}(x), F_{nk}(y)) > \varepsilon$ proving that ε is an expansive constant for F.

Example 2.2 Let N be any positive integer. Consider the time varying map $F = \{f_n\}_{n=0}^{\infty}$ on the unit circle S^1 defined by

$$f_n(z) = \begin{cases} z^{k+1} & 0 \leq n = 2k \leq 2N; \\ z^{\frac{1}{k+2}} & 1 \leq n = 2k+1 < 2N; \\ z & n > 2N. \end{cases}$$

for any $z \in S^1$, where $z^{\frac{1}{m}} = \exp\{\frac{i}{m} Arg(z)\}$, in which $Arg(z)$ is the principal argument of z.

Note that $\{f_n\}_{n=0}^{\infty}$ is equicontinuous family of maps on compact space S^1 and $F^2 = \{g_n\}_{n=0}^{\infty}$, where each $g_n = F_{[2n-1,2n]}$ is the identity map. Since F^2 is not expansive, by Theorem 2.2, F is not expansive.

Example 2.3 Consider the time varying map $F = \{f_n\}_{n=0}^{\infty}$ on the unit circle S^1 defined by

$$f_n(z) = \begin{cases} z^{\frac{n}{2}+1} & \text{if } n \text{ is even;} \\ z^{\frac{2}{n+1}} & \text{if } n \text{ is odd.} \end{cases}$$

for any $z \in S^1$, where $z \in S^1, z^{\frac{1}{m}} = \exp\{\frac{i}{m} Arg(z)\}$, in which $Arg(z)$ is the principal argument of z.

Note that

$$F_n(z) = \begin{cases} z^{\frac{n}{2}+1} & \text{if } n \text{ is even;} \\ z & \text{if } n \text{ is odd.} \end{cases}$$

for any $z \in S^1$ and therefore F is expansive. Observe that $F^2 = \{g_n\}_{n=0}^{\infty}$, where each $g_n = F_{[2n-1,2n]}$ is the identity map, therefore F^2 is not expansive. Note that $\{f_n\}$ is not equicontinuous family on S^1. Thus 'equicontinuity' in the hypothesis of Theorem 2.2 is necessary.

Definition 2.5 Let (X, d) be a metric space, $F = \{f_n\}_{n=0}^{\infty}$ be a time varying map on X and Y be a subset of X. Then Y is said to be invariant under F if $f_n(Y) \subset Y$ for all $n \geq 0$, equivalently $F_n(Y) \subset Y$ for all $n \geq 0$.

Theorem 2.3 *Let (X, d) be a metric space, $F = \{f_n\}_{n=0}^{\infty}$ be a time varying map which is expansive on X and Y be an invariant subset of X, then restriction of F to Y, defined by $F|Y = \{f_n|Y\}$ is expansive.*

Proof Let $\varepsilon > 0$ be an expansive constant for F on X. Let $x \neq y$, $x, y \in Y$ then $x, y \in X$ also, therefore there exists $n \geq 0$ such that $d(F_n(x), F_n(y)) > \varepsilon$. Since Y is invariant under F, $F_n(x), F_n(Y) \in Y$. Hence $F|Y$ is also expansive with expansive constant ε.

Theorem 2.4 *Let (X, d_1) and (Y, d_2) be metric space and $F = \{f_n\}_{n=0}^{\infty}$, $G = \{g_n\}_{n=0}^{\infty}$ be expansive time varying maps on X and Y respectively. Then under metric d on $X \times Y$ defined by*

$$d((x_1, y_1), (x_2, y_2)) = \max\{d_1(x_1, x_2), d_2(y_1, y_2)\}; \quad (x_1, y_1), (x_2, y_2) \in X \times Y,$$

time varying map $F \times G = \{f_n \times g_n\}_{n=0}^{\infty}$ is expansive on $X \times Y$. Hence every finite direct product of expansive time varying maps is expansive.

Proof Note that for any $n \geq 0$,

$$(F \times G)_n(x, y) = (F_n(x), G_n(y)), \quad (x, y) \in X \times Y.$$

Let ε_1 and ε_2 be expansive constants for F and G respectively. Let $\varepsilon = \min\{\varepsilon_1, \varepsilon_2\}$ and $(x_1, y_1), (x_2, y_2) \in X \times Y$.

If for all $n \geq 0$, $d((F \times G)_n(x_1, y_1), (F \times G)_n(x_2, y_2)) < \varepsilon$ then

$$d((F_n(x_1), G_n(y_1)), (F_n(x_2), G_n(y_2))) < \varepsilon$$

which implies

$$\max\{d_1(F_n(x_1), F_n(x_2)), d_2(G_n(y_1), G_n(y_2))\} < \varepsilon.$$

Hence $d_1(F_n(x_1), F_n(x_2)) < \varepsilon \leq \varepsilon_1$ and $d_2(G_n(y_1), G_n(y_2)) < \varepsilon \leq \varepsilon_2$, for all $n \geq 0$ which by expansiveness of F and G implies $x_1 = x_2$ and $y_1 = y_2$. i.e. $(x_1, y_1) = (x_2, y_2)$. Hence $F \times G$ is expansive with expansive constant ε.

3 Shadowing or pseudo orbit tracing property (P.O.T.P.)

Definition 3.1 Let (X, d) be a metric space and $F = \{f_n\}_{n=0}^{\infty}$ be a time varying map on X. For $\delta > 0$, the sequence $\{x_n\}_{n=0}^{\infty}$ in X is said to be a δ-pseudo orbit of F if $d(f_{n+1}(x_n), x_{n+1}) < \delta$ for all $n = 0, 1, 2, \ldots$.

For given $\varepsilon > 0$, a δ-pseudo orbit $\{x_n\}_{n=0}^{\infty}$ is said to be ε-traced by $y \in X$ if $d(F_n(y), x_n) < \varepsilon$ for all $n = 0, 1, 2, \ldots$

The time varying map F is said to have **shadowing property or pseudo orbit tracing property** (P.O.T.P) if for every $\varepsilon > 0$, there exists a $\delta > 0$ such that every δ-pseudo orbit is ε- traced by some point of X.

Remark 3.1 If in the above definition $f_n = f$ for all $n \geq 0$, where $f : X \to X$ is continuous, then P.O.T.P of time varying map $F = \{f_n\}_{n=0}^{\infty}$ on X is equivalent to P.O.T.P. of f on X [2].

Remark 3.2 Note that shadowing property is independent of metric if X is compact. Let d_1 and d_2 be two equivalent metrics on a compact space X. Suppose F has P.O.T.P. in (X, d_1). Let $\varepsilon > 0$ be given. Since d_1 is equivalent to d_2, there exists an $\varepsilon_1 > 0$ such that for any $x \in X$, $N_{d_1}(x, \varepsilon_1) \subset N_{d_2}(x, \varepsilon)$. Since X is compact ε_1 only depends on ε but not on point x. Since F has P.O.T.P. in (X, d_1), for this ε_1, we get a $\delta_1 > 0$ such that any δ_1-pseudo-orbit is ε_1-traced by some point.

Further, since d_1 and d_2 are equivalent, for this δ_1, we get a $\delta > 0$ such that for any $x \in X$, $N_{d_2}(x, \delta) \subset N_{d_1}(x, \delta_1)$. Now let $\{x_n\}_{n=0}^{\infty}$ be a δ-pseudo-orbit of F in (X, d_2). Thus $f_{n+1}(x_n) \in N_{d_2}(x_{n+1}, \delta) \subset N_{d_1}(x_{n+1}, \delta_1)$. Hence $\{x_n\}_{n=0}^{\infty}$ is a δ_1-pseudo orbit of F in (X, d_1), therefore there exists a $y \in X$ which ε_1-traces $\{x_n\}_{n=0}^{\infty}$. Thus $F_n(y) \in N_{d_1}(x_n, \varepsilon_1) \subset N_{d_2}(x_n, \varepsilon)$ which implies $\{x_n\}_{n=0}^{\infty}$ is ε-traced by y in (X, d_2). Hence F has P.O.T.P in (X, d_2).

Theorem 3.1 *Let (X, d_1) and (Y, d_2) be metric spaces. Let $F = \{f_n\}_{n=0}^{\infty}$ and $G = \{g_n\}_{n=0}^{\infty}$ be time varying maps on X and Y respectively such that F is uniformly conjugate to G. If F has P.O.T.P. then G has P.O.T.P.*

Proof Let $\varepsilon > 0$ be given. Since F is uniformly conjugate to G, there exists a uniform homeomorphism $h : X \to Y$ such that $h \circ f_n = g_n \circ h$. i.e. $f_n \circ h^{-1} = h^{-1} \circ g_n$ for all $n \geq 0$. Now h is uniformly continuous being uniform homeomorphism, therefore there exists an $\varepsilon_0 > 0$ such that $d_1(x, y) < \varepsilon_0$ implies $d_2(h(x), h(y)) < \varepsilon$. Since F has P.O.T.P. there exists a $\delta_0 > 0$ such that any δ_0-pseudo orbit of F is ε_0-traced by F orbit of some point of X. Since h being a uniform homeomorphism, h^{-1} is uniformly continuous map, therefore for $\delta_0 > 0$ there exists a $\delta > 0$ such that $d_2(x, y) < \delta$ implies $d_1(h^{-1}(x), h^{-1}(y)) < \delta_0$. Let $\{x_n\}_{n=0}^{\infty}$ be a δ-pseudo orbit for G. i.e. $d_2(g_{n+1}(x_n), x_{n+1}) < \delta$. i.e. $d_1(h^{-1}(g_{n+1}(x_n)), h^{-1}(x_{n+1})) < \delta_0$. i.e. $d_1(f_{n+1}(h^{-1}(x_n)), h^{-1}(x_{n+1})) < \delta_0$ which implies $\{h^{-1}(x_n)\}_{n=0}^{\infty}$ is a δ_0-pseudo orbit for F. Thus there exists a $y \in X$ such that $d_1(F_n(y), h^{-1}(x_n)) < \varepsilon_0$ and hence $d_2(h(F_n(y)), x_n) < \varepsilon$. Now for all $n \geq 0$,

$$h \circ F_n = h \circ f_n \circ f_{n-1} \circ \cdots f_2 \circ f_1 \circ f_0$$
$$= g_n \circ h \circ f_{n-1} \circ \cdots f_2 \circ f_1 \circ f_0$$
$$\vdots$$
$$= g_n \circ g_{n-1} \circ \cdots g_2 \circ g_1 \circ g_0 \circ h$$
$$= G_n \circ h$$

implies $d_2(G_n(h(y)), x_n) < \varepsilon$. i.e. $\{x_n\}_{n=0}^{\infty}$ is ε-traced by $h(y) \in Y$. Thus G has P.O.T.P.

Theorem 3.2 *Let (X, d_1) and (Y, d_2) be metric spaces and $F = \{f_n\}_{n=0}^{\infty}$, $G = \{g_n\}_{n=0}^{\infty}$ be time varying maps on X and Y respectively. Define metric d on $X \times Y$ by*

$$d((x_1, y_1), (x_2, y_2)) = \max\{d_1(x_1, x_2), d_2(y_1, y_2)\}, \quad (x_1, y_1), (x_2, y_2) \in X \times Y.$$

If F and G have P.O.T.P. then the time varying map $F \times G = \{f_n \times g_n\}_{n=0}^{\infty}$ *has P.O.T.P. in* $X \times Y$. *Hence every finite direct product of time varying maps having P.O.T.P., has P.O.T.P.*

Proof Note that for any $n \geq 0$,

$$(F \times G)_n(x, y) = (F_n(x), G_n(y)) \quad (x, y) \in X \times Y.$$

Let $\varepsilon > 0$ be given. Then there exists a $\delta_1 > 0$ and a $\delta_2 > 0$ such that every δ_1-pseudo orbit of F and δ_2-pseudo orbit of G can be ε-traced by some F-orbit and G-orbit respectively. Let $\delta = \min\{\delta_1, \delta_2\}$ and $\{(x_i, y_i)\}_{i=0}^{\infty}$ be a δ pseudo orbit of $F \times G$. Then $d((f_{i+1} \times g_{i+1})(x_i, y_i), (x_{i+1}, y_{i+1})) < \delta$. i.e. $d((f_{i+1}(x_i), g_{i+1}(y_i)), (x_{i+1}, y_{i+1})) < \delta$ which by definition of d implies $d_1(f_{i+1}(x_i), x_{i+1}) < \delta \leq \delta_1$ and $d_2(g_{i+1}(y_i), y_{i+1}) < \delta \leq \delta_2$. Hence there exist $x \in X$ and $y \in Y$ such that $d_1(F_i(x), x_i) < \varepsilon$ and $d_2(G_i(y), y_i) < \varepsilon$. Hence $d((F_i(x), G_i(y)), (x_i, y_i)) < \varepsilon$. i.e. $d((F \times G)_i(x, y), (x_i, y_i)) < \varepsilon$ which implies $\{(x_i, y_i)\}_{i=0}^{\infty}$ is ε-traced by $(x, y) \in X \times Y$. Hence any δ-pseudo orbit of $F \times G$ can be ε-traced by some point of $X \times Y$. Thus $F \times G$ also has P.O.T.P. By induction, we get that every finite direct product of time varying maps having P.O.T.P. has P.O.T.P.

Theorem 3.3 *Let* $F = \{f_n\}_{n=0}^{\infty}$ *be the time varying map on a metric space* (X, d). *If* F *has P.O.T.P. then* F^k *has P.O.T.P. for every* $k > 0$.

Proof If $k = 1$ nothing to prove. Suppose $k \geq 2$. Let $\varepsilon > 0$ be given. Since F has P.O.T.P., therefore there exists a $\delta > 0$ such that every δ-pseudo orbit of F is ε-traced by some point of X. Let $\{y_i\}_{i=0}^{\infty}$ be a δ-pseudo orbit of F^k. Then $d(g_{n+1}(y_n), y_{n+1}) < \delta$ for all $n \geq 0$, where $g_n = F_{[(n-1)k+1, nk]}$, i.e $d(F_{[nk+1, (n+1)k]}(y_n), y_{n+1}) < \delta$, for all $n \geq 0$.

For $0 \leq j < k$ and $n \geq 0$ put $x_{nk+j} = F_{[nk+1, nk+j]}(y_n)$.

Claim $\{x_i\}_{n=0}^{\infty}$ is a δ-pseudo orbit for F.

i.e. to show : $d(f_{nk+j+1}(x_{nk+j}), x_{nk+j+1}) < \delta$, for all $n \geq 0$ and for all j, $0 \leq j < k$.

Choose any $n \geq 0$. Now for any j, $0 \leq j \leq k - 2$,

$$f_{nk+j+1}(x_{nk+j}) = f_{nk+j+1}(F_{[nk+1, nk+j]}(y_n)) = F_{[nk+1, nk+j+1]}(y_n) = x_{nk+j+1}.$$

Thus $d(f_{nk+j+1}(x_{nk+j}), x_{nk+j+1}) = 0 < \delta$ for all j, $0 \leq j \leq k - 2$.

Now for $j = k - 1$,

$$\begin{aligned} d(f_{nk+k}(x_{nk+k-1}), x_{nk+k}) &= d(f_{nk+k}(F_{[nk+1, nk+k-1]}(y_n)), x_{(n+1)k}) \\ &= d(F_{[nk+1, (n+1)k]}(y_n), y_{n+1}) \\ &< \delta \end{aligned}$$

Hence the claim. By P.O.T.P. of F, $\{x_i\}_{i=0}^{\infty}$ is ε-traced by some $y \in X$. i.e. $d(F_i(y), x_i) < \varepsilon$, for all $i \geq 0$. In particular for $i = kn$, $d(F_{kn}(y), x_{kn}) < \varepsilon$. Thus $d(G_n(y), y_n) < \varepsilon$, where $G_n = g_n \circ \cdots \circ g_1 \circ g_0 = F_{nk}$. Thus $F^k = \{g_n\}_{n=0}^{\infty}$ has P.O.T.P.

Remark 3.3 For time varying map given in Example 2.2, $F^2 = \{g_n\}_{n=0}^{\infty}$, where each g_n is the identity map. Now since F^2 does not have P.O.T.P., by above theorem F does not have P.O.T.P.

Lemma 3.1 *Let (X, d) be a compact metric space, $F = \{f_n\}_{n=0}^{\infty}$ be a time varying map on X (where each f_n is continuous on X) and N be a natural number. Then for every $\varepsilon > 0$ there exists $\delta > 0$ such that each δ-finite pseudo orbit $\{x_i : 0 \le i \le N\}$ satisfies $d(F_i(x_0), x_i) < \varepsilon$, for $0 \le i \le N$.*

Proof Let $\varepsilon > 0$ be given. For $N = 1$, take $\delta = \varepsilon$. Here $d(F_0(x_0), x_0) = d(x_0, x_0) = 0 < \varepsilon$ and since $\{x_i : 0 \le i \le 1\}$ is an ε- pseudo orbit, we have $d(f_1(x_0), x_1) < \varepsilon$. i.e. $d(F_1(x_0), x_1) < \varepsilon$. Thus result holds for $N = 1$. Suppose result holds for $N - 1$. Since f_N is uniformly continuous, for every $\varepsilon > 0$ there exists $\varepsilon_1, 0 < \varepsilon_1 < \varepsilon$ such that $d(x, y) < \varepsilon_1$ implies $d(f_N(x), f_N(y)) < \frac{\varepsilon}{2}; x, y \in X$. By our assumption there exists a $\delta_1, 0 < \delta_1 < \varepsilon$ such that each δ_1-pseudo orbit $\{y_i : 0 \le i \le N-1\}$ is ε_1-traced by $y_0 \in X$. We show that each $\frac{\delta_1}{2}$-pseudo orbit $\{x_i : 0 \le i \le N\}$ is ε-traced by $x_0 \in X$.

Since the $\frac{\delta_1}{2}$-pseudo orbit is a δ_1-pseudo-orbit, the finite pseudo-orbit $\{x_i : 0 \le i \le N - 1\}$ is ε_1-traced by the point $x_0 \in X$. Hence $d(F_i(x_0), x_i) < \varepsilon_1 < \varepsilon$, for $0 \le i \le N-1$. In particular, $d(F_{N-1}(x_0), x_{N-1}) < \varepsilon_1$ and so $d(f_N(F_{N-1}(x_0)), f_N(x_{N-1})) < \frac{\varepsilon}{2}$. i.e $d(F_N(x_0), f_N(x_{N-1})) < \frac{\varepsilon}{2}$. Since $\{x_i : 0 \le i \le N\}$ is a $\frac{\delta_1}{2}$-pseudo orbit, we have $d(f_N(x_{N-1}), x_N) < \frac{\delta_1}{2} < \frac{\varepsilon}{2}$ and therefore

$$d(F_N(x_0), x_N) \le d(F_N(x_0), f_N(x_{N-1})) + d(f_N(x_{N-1}), x_N)$$
$$< \frac{\varepsilon}{2} + \frac{\varepsilon}{2}$$
$$= \varepsilon.$$

Hence result is true for N. Thus by principle of mathematical induction the result is true for any natural number N.

For any $m > 0$, replacing f_i by f_{m+i} and x_i by x_{m+i} in Lemma 3.1 we get the following result:

Theorem 3.4 *Let (X, d) be a compact metric space and $F = \{f_n\}_{n=0}^{\infty}$ be a time varying map on X, where family $\{f_n\}$ is equicontinuous on X and N is a natural number. Then for every $\varepsilon > 0$ there exists a $\delta > 0$ such that for each $m \ge 0$ and each δ-pseudo-orbit $\{x_n\}_{n=0}^{\infty}$, the finite δ-pseudo orbit $\{x_{m+i}\}_{i=0}^{N}$ satisfies $d(F_{[m+1,m+i]}(x_m), x_{m+i}) < \varepsilon$ for all i, $0 \le i \le N$.*

Note that since $\{f_n\}$ is equicontinuous δ does not depend upon m.

Theorem 3.5 *Let (X, d) be a compact metric space, $F = \{f_n\}_{n=0}^{\infty}$ be a time varying map on X, where $\{f_n\}_{n=0}^{\infty}$ is equicontinuous on X and $k > 0$. If $F^k = \{g_n\}_{n=0}^{\infty}$, where $g_n = F_{[(n-1)k+1, nk]}$ for all $n \ge 0$, has P.O.T.P. then so does $F = \{f_n\}_{n=0}^{\infty}$.*

Proof Let us first observe the following :

(I) Let $\varepsilon > 0$ be given. Then for each $i \ge 0$, $\{F_{[ik+1, ik+j]} : 0 \le j \le k\}$ is equicontinuous being finite family of continuous functions on compact metric space X therefore there exists an $\varepsilon_1, 0 < \varepsilon_1 < \varepsilon$ such that

$$d(x, y) < \varepsilon_1 \text{ implies } \max_{0 \leq j \leq k} \{d(F_{[ik+1,ik+j]}(x), F_{[ik+1,ik+j]}(y))\} < \frac{\varepsilon}{2}.$$

Note that ε_1 does not depend on i as family $\{f_n\}$ is equicontinuous.

(II) Let ε and ε_1 be as above. Then there exists a δ_0, $0 < \delta_0 < \varepsilon_1$ such that each finite δ_0-pseudo orbit $\{x_{m+j}\}_{j=1}^{k}$ satisfies $d(F_{[m+1,m+j]}(x_m), x_{m+j}) < \frac{\varepsilon}{2}$ for all j, $0 \leq j \leq k$ and for each $m \geq 0$ (by Theorem 3.4).

(III) Let ε_1 and δ_0 be as above. Since F^k has P.O.T.P., there exists a δ_1, $0 < \delta_1 < \delta_0$ such that any δ_1-pseudo-orbit for F^k is ε_1-traced by some point of X.

(IV) Let ε_1 and δ_1 be as above. Then there exists a δ, $0 < \delta < \delta_1$ such that each δ-finite pseudo-orbit $\{z_{m+i}\}_{i=1}^{k}$ satisfies $d(F_{[m+1,m+i]}(z_m), z_{m+i}) < \delta_1$ for all i, $0 \leq i \leq k$ and for each $m \geq 0$ (by Theorem 3.4).

With these properties we prove that each δ-pseudo orbit $\{y_i\}_{i=0}^{\infty}$ of F is ε-traced by some point. Write $x_i = y_{ik}$ for $i = 0, 1, 2, \ldots$ and fix i. Since $\{y_{ik+j} : 0 \leq j \leq k\}$ is a δ-finite pseudo-orbit for F, by (IV) we have $d(F_{[ik+1,ik+j]}(y_{ik}), y_{ik+j}) < \delta_1$ $(0 \leq j \leq k)$ and especially if $j = k$ then $d(F_{[ik+1,ik+k]}(y_{ik}), y_{ik+k}) < \delta_1$ i.e. $d(g_{i+1}(x_i), x_{i+1}) < \delta_1$. i.e. $\{x_i\}_{i=0}^{\infty}$ is δ_1-pseudo orbit for F^k. Hence by (III), there exists $y \in X$ with $d(G_i(y), x_i) < \varepsilon_1$, for $i = 0, 1, 2, \ldots$, where $G_i = g_i \circ g_{i-1} \circ \cdots \circ g_1 \circ g_0 = F_{ik}$. i.e. $d(F_{ik}(y), y_{ik}) < \varepsilon_1$, for $i = 0, 1, 2, \ldots$ and hence from (I) $d(F_{[ik+1,ik+j]}(F_{ik}(y)), F_{[ik+1,ik+j]}(y_{ik})) < \frac{\varepsilon}{2}$. i.e.

$$d(F_{ik+j}(y), F_{[ik+1,ik+j]}(y_{ik})) < \frac{\varepsilon}{2}, \quad o \leq j \leq k, \quad i \geq 0. \tag{1}$$

On the other hand, since $\{y_{ik+j} : 0 \leq j \leq k\}$ is a finite δ_0-pseudo orbit of F (as $\{y_i\}_{i=0}^{\infty}$ is a δ-pseudo orbit and $\delta < \delta_1 < \delta_0$, so $\{y_i\}_{i=0}^{\infty}$ is also δ_0-pseudo orbit of F), from (II) it follows that for any $i \geq 0$, taking $m = ik$,

$$d(F_{[ik+1,ik+j]}(y_{ik}), y_{ik+j}) < \frac{\varepsilon}{2}, \tag{2}$$

for $0 \leq j \leq k$. Therefore using Eqs. (1) and (2), we get

$$\begin{aligned} d(F_{ik+j}(y), y_{ik+j}) &\leq d(F_{ik+j}(y), F_{[ik+1,ik+j]}(y_{ik})) + d(F_{[ik+1,ik+j]}(y_{ik}), y_{ik+j}) \\ &< \frac{\varepsilon}{2} + \frac{\varepsilon}{2} \\ &= \varepsilon, \end{aligned}$$

for $0 \leq j \leq k$. Since i is arbitrary, we have $d(F_n(y), y_n) < \varepsilon$ for $n = 0, 1, 2 \ldots$ and hence the δ-pseudo orbit $\{y_n\}_{n=0}^{\infty}$ is ε-traced by y.

Example 3.1 Let $M > 0$ and $F = \{f_n\}_{n=0}^{\infty}$ be a time varying map on $[0, 1]$ defined by

$$f_n = \begin{cases} x^{2(k+1)} & 0 \leq n = 2k \leq 2M \\ x^{\frac{1}{k+2}} & 1 \leq n = 2k+1 < 2M \\ x^2 & n > 2M. \end{cases}$$

Here $F^2 = \{g_n\}_{n=0}^{\infty}$, where $g_n(x) = x^2$ for all $n \geq 0$. Since map x^2 on $[0, 1]$ has P.O.T.P. (by Lemma 4.1 in [6]), F^2 has P.O.T.P. By above theorem we get that F has P.O.T.P.

4 Topological stability of a time varying map on a compact metric space

Let (X, d) be a compact metric space, define standard bounded metric d_1 on X by

$$d_1(x, y) = \max\{d(x, y), \ 1\}, \quad x, y \in X.$$

and $(\mathcal{C}(X), \eta)$ be the space of all continuous self maps on X, where metric η is defined by

$$\eta(f, g) = \sup_{x \in X} d_1(f(x), g(x)), \quad f, g \in \mathcal{C}(X)$$

A time varying map is a countable subset of $\mathcal{C}(X)$. Let $S(X)$ be the collection of all time varying maps on X. We define a metric ρ on $S(X)$ as follows:

For $F = \{f_n\}_{n=0}^{\infty}$ and $G = \{g_n\}_{n=0}^{\infty}$,

$$\rho(F, G) = \sup_{n \geq 0} \eta(f_n, g_n).$$

Definition 4.1 A time varying map F is said to be topologically stable in $S(X)$, if for every $\varepsilon > 0$ there exists a $0 < \delta < 1$ such that for a time varying map G with $\rho(F, G) < \delta$ there is a continuous map h so that for all $x \in X$, $d(h(x), x) < \varepsilon$ and $d(F_n(h(x)), G_n(x)) < \varepsilon$, for all $n \geq 0$.

Theorem 4.1 *If a time varying map F on a compact metric space X is expansive and has P.O.T.P. then it is topologically stable in $S(X)$.*

Proof Let $e > 0$ be an expansive constant for time varying map $F = \{f_n\}_{n=0}^{\infty}$ and fix $0 < \varepsilon < \frac{e}{3}$. Let $0 < \delta < \min\{\frac{e}{3}, 1\}$ be such that every δ-pseudo orbit of F can be ε-traced by some F-orbit. By expansiveness of F, it follows that there exists a unique $x \in X$ which ε-traces a given δ-pseudo orbit $\{x_i\}_{i=0}^{\infty}$. Indeed, let $y \in X$ be another ε-tracing point of $\{x_i\}_{i=0}^{\infty}$. Then we have $d(F_n(x), x_n) < \varepsilon$ and $d(F_n(y), x_n) < \varepsilon$ for all $n \geq 0$. Now

$$d(F_n(x), F_n(y)) \leq d(F_n(x), x_n) + d(x_n, F_n(y)) \leq 2\varepsilon < e,$$

for all $n \geq 0$ and hence $x = y$.

Let $G = \{g_n\}_{n=0}^{\infty}$ be a time varying map on X such that $\rho(F, G) < \delta$. Since $\delta < 1$, $d(f_n(x), g_n(x)) < \delta$, for all $x \in X$ and for all $n \geq 0$. Let $x \in X$. Since $d(f_{n+1}(G_n(x)), G_{n+1}(x)) = d(f_{n+1}(G_n(x)), g_{n+1}(G_n(x))) < \delta$, for all $n \geq 0$, $\{G_n(x)\}_{n=0}^{\infty}$ is a δ pseudo orbit for $F = \{f_n\}_{n=0}^{\infty}$. Thus there exists a unique point $h(x) \in X$ whose F-orbit ε-traces $\{G_n(x)\}_{n=0}^{\infty}$. This defines a map $h \colon X \to X$ with $d(F_n(h(x)), G_n(x)) < \varepsilon$ for $n \geq 0$ and $x \in X$. Letting $n = 0$, we have $d(h(x), x) < \varepsilon$, for each $x \in X$.

Finally, we show that h is continuous. Let $\lambda > 0$. Then we can choose $N > 0$ such that,

$$0 \leq n \leq N, \quad d(F_n(x), F_n(y)) < e \Rightarrow d(x, y) < \lambda. \tag{3}$$

Suppose this is false. Let α be an open cover of X with diameter less than e. Then there exists an $\varepsilon > 0$ such that for each $j \geq 0$, there exist $x_j, y_j \in X$ with $d(x_j, y_j) > \varepsilon$ and $A_{j,i} \in \alpha$ $(0 \leq i \leq j)$ with $F_i(x_j), F_i(y_j) \in A_{j,i}$, $0 \leq i \leq j$. Since X is compact, there exist $x, y \in X$ such that $x_j \to x$ and $y_j \to y$. Note that $x \neq y$ as for each $j, d(x_j, y_j) > \varepsilon$. Consider the sets $A_{j,0}$ for $j \geq 0$. Since X is compact therefore α has finite subcover and hence infinitely many $A_{j,0}$ coincide to some $A_0 \in \alpha$. Thus $x_j, y_j \in A_0$ for infinitely many j which implies $x, y \in \bar{A}_0$. Similarly for infinitely many $A_{j,n}$ they coincide to some $A_n \in \alpha$ and we have $F_n(x_j), F_n(y_j) \in A_n$ for infinitely many j. Thus F_n being continuous, $F_n(x), F_n(y) \in A_n$. Since $diam(A_n) = diam(\bar{A}_n) < e$, we have $d(F_n(x), F_n(y)) < e$ for all $n \geq 0$, contradicting the fact that F is expansive.

Since $\{G_n : 0 \leq n \leq N\}$ is uniformly equicontinuous on X, we can choose $\eta > 0$ such that $d(x, y) < \eta$ implies $d(G_n(x), G_n(y)) < \frac{e}{3}$ for all $n, 0 \leq n \leq N$.

If $d(x, y) < \eta$ then for all $n, 0 \leq n \leq N$, we have

$$d(F_n(h(x)), F_n(h(y))) \leq d(F_n(h(x)), G_n(x)) + d(G_n(x), G_n(y))$$
$$+ d(G_n(y), F_n(h(y))) \leq \varepsilon + \frac{e}{3} + \varepsilon$$
$$< e.$$

Thus by Eq. (3), $d(h(x), h(y)) < \lambda$. Therefore $d(x, y) < \eta$ implies $d(h(x), h(y)) < \lambda$ proving continuity of h.

Remark 4.1 We generalize the concept of expansiveness for invertible time varying dynamical systems. If we consider the sequence $\{f_n\}_{n=0}^{\infty}$ as the time varying map F, where each f_n is a homeomorphism on a compact metric space X with f_0 as the identity map on X, we define for $n \geq 0$,

$$F_n = f_n \circ f_{n-1} \circ \cdots \circ f_1 \circ f_0$$

and for $n = -m < 0$,

$$F_n = F_{(-m)} = (F_m)^{-1} = f_0^{-1} \circ f_1^{-1} \circ \cdots \circ f_{(m-1)}^{-1} \circ f_m^{-1}.$$

For $x_0 \in X$, orbit of x_0 is the sequence $\{x_n\}_{n=-\infty}^{\infty}$ where $x_{n+1} = f_{n+1}(x_n)$ for all $n \geq 0$ and $x_n = f_{(-n)}^{-1}(x_{n+1})$ for all $n < 0$.

For invertible time varying dynamical system F, we say that F is expansive if there exists a constant $c > 0$ (called an expansive constant) such that for any $x, y \in X$, $x \neq y, d(F_n(x), F_n(y)) > c$ for some integer n.

Analogously we can define shadowing property and stability of invertible time varying dynamical systems and prove related results for such systems which are proved for classical dynamical systems given by a homeomorphism.

Acknowledgments Authors thank the referee for drawing their attention to reference [1] and for some useful suggestions.

References

1. Araújo, V.: Attractors and time averages for random maps. Ann. Inst. H. Poincaré Anal. Non Linéaire **17**(3), 307–369 (2000)
2. Aoki, N., Hiraide, K.: Topological Theory of Dynamical Systems. North-Holland Publishing Co., Amsterdam (1994)
3. Bowen, R.: Entropy-expansive maps. Trans. Am. Math. Soc. **164**, 323–331 (1972)
4. Bessa, M., Rocha, J.: A remark on the topological stability of symplectomorphisms. Appl. Math. Lett. **25**, 163–165 (2012)
5. Bessa, M., Rocha, J.: Topological stability for conservative systems. J. Differ. Equ. **250**(10), 3960–3966 (2011)
6. Chen, L., Li, S.: Shadowing property for inverse limit spaces. Proc. Am. Math. Soc. **115**(2), 573–580 (1992)
7. Das, R.: Chaos of a sequence of maps in a metric G-space. Appl. Math. Sci. **136**(6), 6769–6775 (2012)
8. Das, T., Lee, K., Richeson, D., Wiseman, J.: Spectral decomposition for topologically Anosov homeomorphisms on noncompact and non-metrizable spaces. Topology Appl. **160**(1), 149–158 (2013)
9. Eisenberg, M.: Expansive transformation semigroups of endomorphisms. Fund. Math. **59**, 313–321 (1966)
10. Hurley, M.: Fixed points of topologically stable flows. Trans. Am. Math. Soc. **294**(2), 625–633 (1986)
11. Hurley, M.: Consequences of topological stability. J. Differ. Equ. **54**(1), 60–72 (1984)
12. Hurley, M.: Combined structural and topological stability are equivalent to Axiom A and the strong transversality condition. Ergodic Theory Dyn. Syst. **4**(1), 81–88 (1984)
13. Kato, H.: Continuum-wise expansive homeomorphisms. Can. J. Math. **45**, 576–598 (1993)
14. Lee, K., Sakai, K.: Various shadowing properties and their equivalence. Discrete Contin. Dyn. Syst. **13**(2), 533–540 (2005)
15. Morales, C.: A generalization of expansivity. Discrete Contin. Dyn. Syst. **32**(1), 293–301 (2012)
16. Moriyasu, K.: The topological stability of diffeomorphisms. Nagoya Math. J. **123**, 91–102 (1991)
17. Morales, C.: Measure-expansive systems. Preprint IMPA, D083 (2011)
18. Moriyasu, K., Sakai, K., Sumi, N.: Vector fields with topological stability. Trans. Am. Math. Soc. **353**(8), 3391–3408 (2001)
19. Nitecki, Z.: On semi-stability for diffeomorphisms. Invent. Math. **14**, 83–122 (1971)
20. Pilyugin, S.Y.: Shadowing in dynamical systems, Lecture Notes in Mathematics, vol. 1706. Springer, Berlin (1999)
21. Reddy, W.L.: Pointwise expansion homeomorphisms. J. Lond. Math. Soc. **2**, 232–236 (1970)
22. Robinson, C.: Stability theorems and hyperbolicity in dynamical systems. Rocky Mt. J. Math. **7**(3), 425–437 (1977)
23. Tian, C., Chen, G.: Chaos of a sequence of maps in a metric space. Chaos Solitons Fractals **28**, 1067–1075 (2006)
24. Utz, W.R.: Unstable homeomorphisms. Proc. Am. Math. Soc. **1**, 769–774 (1950)
25. Walters, P.: Anosov diffeomorphism are topologically stable. Topology **9**, 71–78 (1970)
26. Walters, P.: On the pseudo orbit tracing property and its relationship to stability, Lecture Notes in Mathematics, vol. 668. Springer, Berlin, pp. 231–244 (1978)

3

Cauchy–Riemann meet Monge–Ampère

Zbigniew Błocki

Abstract This is a relatively self-contained introduction to recent developments in the $\bar\partial$-equation, Ohsawa–Takegoshi extension theorem and applications of pluripotential theory to the Bergman kernel and metric. The main tools are the Hörmander L^2-estimate for $\bar\partial$ and Bedford–Taylor's theory of the complex Monge–Ampère operator.

Contents

Communicated by Neil Trudinger.

Partially supported by the Ideas Plus grant 0001/ID3/2014/63 of the Polish Ministry of Science and Higher Education.

Z. Błocki (✉)
Instytut Matematyki, Uniwersytet Jagielloński, Łojasiewicza 6, 30-348 Krakow, Poland
e-mail: zbigniew.blocki@im.uj.edu.pl; umblocki@cyf-kr.edu.pl

1 Introduction

Holomorphic functions of several variables are precisely solutions to the homogeneous Cauchy–Riemann equation (often called the $\bar{\partial}$-equation)

$$\bar{\partial}u = 0. \tag{1.1}$$

Here both sides are forms of type $(0,1)$ which is a rather special case of the $\bar{\partial}$-equation because all solutions, even in the distributional sense, have to be smooth, in contrast to the general case of the equation for (p, q)-forms. The inhomogeneous $\bar{\partial}$-equation

$$\bar{\partial}u = \alpha, \tag{1.2}$$

where α is a $\bar{\partial}$-closed $(0,1)$-form, plays a fundamental role in the PDE approach to the theory of several complex variables: it is the main tool for constructing holomorphic functions. The basic idea is very simple: if $\alpha = \bar{\partial}\chi$ for some function χ and u is a solution to (1.2) then $u - \chi$ is holomorphic.

The famous L^2-estimate of Hörmander [65] asserts that for every smooth strongly plurisubharmonic function φ defined in a pseudoconvex open subset of $\mathbb{C}^n$ there exists a solution to (1.2) satisfying

$$\int_\Omega |u|^2 e^{-\varphi} d\lambda \le \int_\Omega |\alpha|^2_{i\partial\bar{\partial}\varphi} e^{-\varphi} d\lambda. \tag{1.3}$$

The original Hörmander estimate was slightly weaker: the right-hand side depended on the minimal eigenvalue of the complex Hessian of φ but his method also gives this slightly stronger version (it was first formulated by Demailly [42]). This turns out to be an extremely powerful result as will be again demonstrated here. What makes this approach so useful is a big abundance of plurisubharmonic functions: they are usually much easier to construct than holomorphic functions and this is in fact where pluripotential theory comes into play.

As we will see, Hörmander's estimate (1.3) can also be formulated for non-smooth φ. In many cases an almost optimal choice for the weight φ in this and related estimates is

$$\varphi = 2nG_\Omega(\cdot, w),$$

where $G_\Omega(\cdot, w)$ is the pluricomplex Green function with pole at w. This is because it is essentially the largest negative plurisubharmonic function such that $e^{-\varphi}$ is not locally integrable near w. This is the main reason why pluripotential theory turned out to be so useful in the theory of the $\bar{\partial}$-equation.

The complex Monge–Ampère operator $(dd^c)^n$ plays the central role in pluripotential theory, it has been developed in this context by Bedford and Taylor [1,2]. For example, Demailly [43] characterized the pluricomplex Green function as a solution to the Monge–Ampère equation with point-mass on the right-hand side. This, together with standard techniques for the complex Monge–Ampère operator, e.g. integrating by parts, is often used to prove various properties of the Green function.

One of the most important results in several complex variables has been the Ohsawa–Takegoshi extension theorem [98]. It states that holomorphic functions can be extended from lower dimensional sections with L^2-estimates. It has found many applications in complex and algebraic geometry but it can be also very useful to study singularities of plurisubharmonic functions. For example, it turns out that two main results in this area, the theorem of Siu [107] on analyticity of level sets of Lelong numbers and the openness conjecture of Demailly and Kollár [46] follow relatively easily from the Ohsawa–Takegoshi theorem. The simple proof of the Siu theorem was found by Demailly [45] who devised a special approximation of an arbitrary plurisubharmonic function by smooth ones with possibly analytic singularities. The openness conjecture was first proved by Berndtsson [10] who subsequently simplified the proof in [11] using an approach of Guan and Zhou [58].

This survey is largely self-contained. It is organized as follows. In Sect. 2 we give proofs of all necessary L^2-estimates for $\bar{\partial}$ assuming Hörmander's estimate. It is mostly thanks to the method of Berndtsson from [5] that they are in fact formal consequences of (1.3) and one does not have to repeat Hörmander's arguments. Section 3 contains the simplest known proof of the Ohsawa–Takegoshi extension theorem. It is due to Chen [39] (see also [26]) and was the first one which used Hörmander's estimate directly. In Sect. 4 we present some applications of the Ohsawa–Takegoshi theorem to singularities of plurisubharmonic functions with simple proofs of the aforementioned openness conjecture and Siu's theorem, as well as basic results on Demailly's approximation. Section 5 is a brief introduction to the complex Monge–Ampère operator and the pluricomplex Green function. Section 6 discusses some applications of pluripotential theory and the $\bar{\partial}$-equation to the Bergman metric. In Sect. 7 we present the recently settled (see [27]) one-dimensional Suita conjecture from [110] and closely related versions of the Ohsawa–Takegoshi theorem with optimal constant. Another approach to the Suita conjecture from [28] and its multidimensional version from [31] are also discussed. The case of convex domains is analysed in greater detail in Sect. 8, following mostly [31], and it is used in Sect. 9 to present recent Nazarov's proof [93] of the Bourgain–Milman inequality [34] from convex analysis. Finally, in Sect. 10 we discuss a link between the lower bound for the Bergman kernel in terms of the pluricomplex Green function and possible symmetrization results for the complex Monge–Ampère equation and complex isoperimetric inequalities. The conclusion of this section is rather speculative in nature. Many open problems are mentioned throughout the whole paper.

A large part of this paper was written during author's stay at the Korea Institute for Advanced Study in Seoul. He is grateful to Mihai Paun and others at KIAS for the invitation, great hospitality and very stimulating atmosphere.

2 L^2-estimates for the $\bar{\partial}$-equation

We first recall the definition of the operator $\bar{\partial}$ for functions and $(0, 1)$-forms (this is all we will need). For a function u defined on an open subset of $\mathbb{C}^n$ set

$$\bar{\partial} u := \sum_j \frac{\partial u}{\partial \bar{z}_j} d\bar{z}_j$$

and for a (0,1)-form $\alpha = \sum_k \alpha_k d\bar{z}_k$

$$\bar{\partial}\alpha =: \sum_k \bar{\partial}\alpha_k \wedge d\bar{z}_k = \sum_{j<k} \left(\frac{\partial \alpha_k}{\partial \bar{z}_j} - \frac{\partial \alpha_j}{\partial \bar{z}_k} \right) d\bar{z}_j \wedge d\bar{z}_k.$$

We will consider the inhomogeneous $\bar{\partial}$-equation

$$\bar{\partial} u = \alpha \qquad\qquad (2.1)$$

which is really a system of n equations with one unknown:

$$\frac{\partial u}{\partial \bar{z}_j} = \alpha_j, \quad j = 1, \ldots, n.$$

Since $\bar{\partial}^2 = 0$, a necessary condition for (2.1) to have a solution is $\bar{\partial}\alpha = 0$, that is

$$\frac{\partial \alpha_j}{\partial \bar{z}_k} = \frac{\partial \alpha_k}{\partial \bar{z}_j}.$$

Recall that a function φ, defined on an open subset of $\mathbb{C}^n$ with values in $[-\infty, \infty)$, is called *plurisubharmonic* if locally it is upper semi-continuous, $\not\equiv -\infty$ and is either subharmonic or $\equiv -\infty$ on every complex line. Equivalently, the complex Hessian $(\partial^2 \varphi / \partial z_j \bar{\partial} z_k)$ is positive semi-definite (in the distributional sense). It is in fact an open problem whether upper semi-continuity in the first definition follows from the other properties. We use the notation $PSH(\Omega)$ for the set of all plurisubharmonic functions in Ω and $PSH^-(\Omega)$ for the negative ones. The C^2 functions with positive definite complex Hessian at every point are called *strongly plurisubharmonic*. An open subset $\Omega \subset \mathbb{C}^n$ is called *pseudoconvex* if it admits a plurisubharmonic exhaustion function, that is there exists $\varphi \in PSH(\Omega)$ such that $\{\varphi \leq t\} \Subset \Omega$ for all $t \in \mathbb{R}$. A C^2 smooth Ω is called *strongly pseudoconvex* if it admits a strongly plurisubharmonic defining function, that is strongly plurisubharmonic ρ defined in a neighbourhood of $\bar{\Omega}$ such that $\nabla \rho \neq 0$ on $\partial\Omega$ and $\Omega = \{\rho < 0\}$.

The notions of plurisubharmonic and strongly plurisubharmonic functions as well as pseudoconvex and strongly pseudoconvex sets in $\mathbb{C}^n$ correspond closely to that of convex and strongly convex functions and domains in $\mathbb{R}^m$. In this context, also the $\bar{\partial}$-operator can be treated as a counterpart of the d-operator, see [7].

We want to formulate Hörmander's estimate (1.3) also for non-smooth φ (see [23]). Notice that if φ is C^2 and strongly plurisubharmonic then

$$H := |\alpha|^2_{i\partial\bar{\partial}\varphi} = \sum_{j,k} \varphi^{j\bar{k}} \bar{\alpha}_j \alpha_k,$$

where $(\varphi^{j\bar{k}}) = (\partial^2\varphi/\partial z_j\bar{\partial}z_k)^{-1}$, is the smallest function satisfying

$$(\bar{\alpha}_j\alpha_k) \leq H(\partial^2\varphi/\partial z_j\bar{\partial}z_k).$$

This can be written as

$$i\bar{\alpha} \wedge \alpha \leq Hi\partial\bar{\partial}\varphi. \tag{2.2}$$

Note that if the coefficients of α are in L^2_{loc} and H is in L^∞_{loc} then both sides of (2.2) are well defined currents of order 0 (that is forms with complex measures as coefficients).

We can now state Hörmander's estimate as follows:

Theorem 2.1 *Assume that Ω is a pseudoconvex open subset of $\mathbb{C}^n$ and $\varphi \in PSH(\Omega)$. Let $\alpha \in L^2_{loc,(0,1)}(\Omega)$ be $\bar{\partial}$-closed and take non-negative $H \in L^\infty_{loc}(\Omega)$ satisfying (2.2). Then there exists $u \in L^2_{loc}(\Omega)$ solving (2.1) and such that*

$$\int_\Omega |u|^2 e^{-\varphi}d\lambda \leq \int_\Omega He^{-\varphi}d\lambda.$$

This estimate is easier to prove for $n = 1$, see [67]. As remarked by Berndtsson [9], it is therefore quite surprising that it had not been proved earlier in this case. Especially that it can lead to new nontrivial results in one dimensional complex analysis, see e.g. [27]. But of course it is especially powerful in higher dimensions. For example the solution of the Levi problem can be deduced quite easily from Theorem 2.1, see [67, Corollary 4.2.8].

Sometimes there is however an inconvenience with applying Hörmander's estimate directly: φ appears both as a weight as well as a Kähler potential on the right-hand side. The following estimate due to Donnelly and Fefferman [52] (formulated originally for $\varphi \equiv 0$) addressed this problem:

Theorem 2.2 *Let Ω, φ and α be as in Theorem 2.1. Assume in addition that $\psi \in PSH(\Omega)$ is such that*

$$i\partial\psi \wedge \bar{\partial}\psi \leq i\partial\bar{\partial}\psi. \tag{2.3}$$

Then there exists $u \in L^2_{loc}(\Omega)$ solving (2.1) and such that

$$\int_\Omega |u|^2 e^{-\varphi}d\lambda \leq C\int_\Omega |\alpha|^2_{i\partial\bar{\partial}\psi}e^{-\varphi}d\lambda$$

for some absolute constant C.

Theorem 2.2 is stated here somewhat imprecisely although it is rather clear what the right statement should be: if ψ is not smooth and strongly plurisubharmonic then $|\alpha|^2_{i\partial\bar{\partial}\psi}$ should be replaced by any non-negative locally bounded H such that $i\bar{\alpha} \wedge \alpha \leq Hi\partial\bar{\partial}\psi$. Plurisubharmonic functions satisfying (2.3) are precisely of the form

$$\psi = -\log(-v)$$

for some $v \in PSH^-(\Omega)$. It was shown by Berndtsson [5] that Theorem 2.2 is a formal consequence of Hörmander's estimate:

Proof of Theorem 2.2 By standard approximation we may assume that ψ is smooth, strongly plurisubharmonic and that Ω, φ, ψ are bounded. Let u be the solution to (2.1) which is minimal in the $L^2(\Omega, e^{-\varphi-\psi/2})$-norm. This means that u is perpendicular to $\ker \bar{\partial}$ in $L^2(\Omega, e^{-\varphi-\psi/2})$, that is

$$\int_\Omega u \bar{f} e^{-\varphi-\psi/2} d\lambda = 0, \quad f \in \ker \bar{\partial},$$

and therefore

$$v := e^{\psi/2} u$$

is the minimal solution to

$$\bar{\partial} v = \beta,$$

where $\beta = e^{\psi/2}\left(\alpha + u \, \bar{\partial}\psi/2\right)$, in the $L^2(\Omega, e^{-\varphi-\psi})$-norm. (Note that by our regularity assumptions the spaces $L^2(\Omega, e^{-\varphi-\psi/2})$ and $L^2(\Omega, e^{-\varphi-\psi})$ are the same as sets and so is $\ker \bar{\partial}$ in both cases.) Theorem 2.1 implies that

$$\int_\Omega |v|^2 e^{-\varphi-\psi} d\lambda \leq \int_\Omega |\beta|^2_{i\partial\bar{\partial}(\varphi+\psi)} e^{-\varphi-\psi} d\lambda \leq \int_\Omega |\beta|^2_{i\partial\bar{\partial}\psi} e^{-\varphi-\psi} d\lambda,$$

that is

$$\int_\Omega |u|^2 e^{-\varphi} d\lambda \leq \int_\Omega |\alpha + u \, \bar{\partial}\psi/2|^2_{i\partial\bar{\partial}\psi} e^{-\varphi} d\lambda.$$

By (2.3) for any $t > 0$

$$|\alpha + u \, \bar{\partial}\psi/2|^2_{i\partial\bar{\partial}\psi} \leq \left(1 + \frac{t}{2}\right)|\alpha|^2_{i\partial\bar{\partial}\psi} + \left(\frac{1}{4} + \frac{1}{2t}\right)|u|^2$$

and we obtain the required estimate if we take any $t > 2/3$, with the optimal choice $t = 2$, we then get $C = 4$. ☐

The idea of *twisting* the $\bar{\partial}$-equation seen in the proof of Theorem 2.2 had been used before but Berndtsson [5] seems to have been the first to realize that it can be applied directly to Hörmander's estimate, without repeating the technical parts of its proof like the so called Bochner–Kodaira identity, integration by parts etc.

The constant $C = 4$ we got here was originally obtained in [21] and it was shown to be optimal in [29]. Take $\Omega = \Delta$, the unit disc in $\mathbb{C}$, $\varphi \equiv 0$ and $\psi(z) = -\log(-\log|z|)$. For smooth, compactly supported η on $(0, \infty)$ one can show that

$$u(z) = \frac{\eta(-\log|z|)}{z}$$

is the minimal solution to (2.1) in $L^2(\Delta)$, where

$$\alpha = -\frac{\eta'(-\log|z|)}{2|z|^2}d\bar{z}.$$

Then by Theorem 2.2

$$\int_0^\infty \eta^2 dt \le 4 \int_0^\infty (\eta')^2 t^2 dt, \quad \eta \in W_0^{1,2}((0,\infty)),$$

and one can show the constant 4 cannot be improved here.

The Donnelly–Fefferman estimate was generalized by Berndtsson [4]: he showed that with the assumptions of Theorem 2.2 and with $0 \le \delta < 1$ one can obtain solution u satisfying

$$\int_\Omega |u|^2 e^{\delta\psi - \varphi} d\lambda \le \frac{4}{(1-\delta)^2} \int_\Omega |\alpha|^2_{i\partial\bar\partial\psi} e^{\delta\psi - \varphi} d\lambda. \tag{2.4}$$

This particular constant was obtained in [21] and, similarly as above, it was shown in [29] to be optimal for every δ.

Berndtsson's estimate is closely related to the Ohsawa–Takegoshi extension theorem, see [4], but the latter cannot be deduced directly from it. If (2.4) were true for $\delta = 1$ (with some finite constant) then it would be sufficient. Building on an idea of Chen [40] in his remarkable proof of the extension theorem, this was overcome in [26]. The following is a counterpart of Berndtsson's estimate (2.4) for $\delta = 1$:

Theorem 2.3 *Let Ω, φ, ψ and α be as in Theorem 2.2. Assume in addition that $|\bar\partial\psi|^2_{i\partial\bar\partial\psi} \le a < 1$ on $\operatorname{supp}\alpha$ (note that (2.2) means that $|\bar\partial\psi|^2_{i\partial\bar\partial\psi} \le 1$ in Ω). Then we can find a solution $u \in L^2_{loc}(\Omega)$ to (2.1) satisfying*

$$\int_\Omega \left(1 - |\bar\partial\psi|^2_{i\partial\bar\partial\psi}\right) |u|^2 e^{\psi - \varphi} d\lambda \le \frac{1+\sqrt{a}}{1-\sqrt{a}} \int_\Omega |\alpha|^2_{i\partial\bar\partial\psi} e^{\psi - \varphi} d\lambda.$$

The trade-off compared with the previous estimates is the extra error term on the left-hand side. On the other hand, this estimate can be used to prove the Ohsawa–Takegoshi theorem directly as we will see in Sect. 3. It is however not sufficient to get the extension theorem with optimal constant. A more general one which is sufficient for that purpose is the following $\bar\partial$-estimate from [27] where only one weight has to be plurisubharmonic and the other one is essentially arbitrary:

Theorem 2.4 *Let Ω be pseudoconvex in $\mathbb{C}^n$ and $\alpha \in L^2_{loc,(0,1)}(\Omega)$ be $\bar\partial$-closed. Assume that $\varphi \in PSH(\Omega)$ and $\psi \in W^{1,2}_{loc}(\Omega)$ which is locally bounded from above satisfy*

$$|\bar\partial\psi|^2_{i\partial\bar\partial\varphi} \le \begin{cases} 1 & in \ \Omega \\ a < 1 & on \ \operatorname{supp}\alpha \end{cases}.$$

Then there exists $u \in L^2_{loc}(\Omega)$ solving (2.1) and such that

$$\int_\Omega \left(1 - |\bar{\partial}\psi|^2_{i\partial\bar\partial\varphi}\right) |u|^2 e^{2\psi-\varphi} d\lambda \leq \frac{1+\sqrt{a}}{1-\sqrt{a}} \int_\Omega |\alpha|^2_{i\partial\bar\partial\varphi} e^{2\psi-\varphi} d\lambda.$$

Proof The proof will be similar to that of Theorem 2.2. Again by approximation we may assume that ψ is smooth, strongly plurisubharmonic and Ω, φ, ψ are bounded. Let u be the minimal solution to (2.1) in $L^2(\Omega, e^{\psi-\varphi})$. Since u is perpendicular to ker $\bar{\partial}$ in $L^2(\Omega, e^{\psi-\varphi})$, it follows that $v := ue^\psi$ is perpendicular to ker $\bar{\partial}$ in $L^2(\Omega, e^{-\varphi})$. Therefore v is the minimal solution to $\bar{\partial}v = \beta := e^\psi(\alpha + u\,\bar{\partial}\psi)$ in $L^2(\Omega, e^{-\varphi})$ and by Hörmander's estimate

$$\int_\Omega |v|^2 e^{-\varphi} d\lambda \leq \int_\Omega |\beta|^2_{i\partial\bar\partial\varphi} e^{-\varphi} d\lambda.$$

Therefore

$$\int_\Omega |u|^2 e^{2\psi-\varphi} d\lambda \leq \int_\Omega |\alpha + u\,\bar{\partial}\psi|^2_{i\partial\bar\partial\varphi} e^{2\psi-\varphi} d\lambda$$
$$\leq \int_\Omega \left(|\alpha|^2_{i\partial\bar\partial\varphi} + 2|u|\sqrt{H}|\alpha|_{i\partial\bar\partial\varphi} + |u|^2 H\right) e^{2\psi-\varphi} d\lambda,$$

where $H = |\bar{\partial}\psi|^2_{i\partial\bar\partial\varphi}$. For $t > 0$ we will get

$$\int_\Omega |u|^2 (1-H) e^{2\psi-\varphi} d\lambda$$
$$\leq \int_\Omega \left[|\alpha|^2_{i\partial\bar\partial\varphi} \left(1 + t^{-1}\frac{H}{1-H}\right) + t|u|^2(1-H)\right] e^{2\psi-\varphi} d\lambda$$
$$\leq \left(1 + t^{-1}\frac{a}{1-a}\right) \int_\Omega |\alpha|^2_{i\partial\bar\partial\varphi} e^{2\psi-\varphi} d\lambda$$
$$+ t \int_\Omega |u|^2(1-H) e^{2\psi-\varphi} d\lambda.$$

We will obtain the required estimate if we take $t = 1/(a^{-1/2} + 1)$.   □

This is the most general $\bar{\partial}$-estimate of all discussed so far. First of all note that, unlike the previous ones, it recovers Hörmander's estimate: it is enough to take $\psi \equiv 0$ and $a = 0$. It also easily implies all the previous results with optimal constants. To obtain Berndtsson's estimate (2.4) (and thus also Donnelly–Fefferman's for $\delta = 0$) for plurisubharmonic φ, ψ satisfying (2.2) and $\delta < 1$ set

$$\tilde{\varphi} := \varphi + \psi, \quad \tilde{\psi} := \frac{1+\delta}{2}\psi.$$

Then $2\tilde{\psi} - \tilde{\varphi} = \delta\psi - \varphi$ and

$$|\bar{\partial}\tilde{\psi}|^2_{i\partial\bar\partial\tilde\varphi} \leq \frac{(1+\delta)^2}{4} =: a.$$

Theorem 2.4 will give (2.4) with the constant

$$\frac{1 + \sqrt{a}}{(1 - \sqrt{a})(1 - a)} = \frac{4}{(1 - \delta)^2}.$$

If φ, ψ and a are as in Theorem 2.3 and $\widetilde{\varphi} := \varphi + \psi$ then $|\bar{\partial}\psi|^2_{i\partial\bar{\partial}\widetilde{\varphi}} \leq |\bar{\partial}\psi|^2_{i\partial\bar{\partial}\psi}$ and Theorem 2.4 immediately gives Theorem 2.3.

3 Ohsawa–Takegoshi extension theorem

The following theorem proved by Ohsawa and Takegoshi [98] turned out to be one of the most important results in complex analysis and complex geometry.

Theorem 3.1 *Let Ω be a bounded pseudoconvex open set in $\mathbb{C}^n$ and let H be an affine complex subspace of $\mathbb{C}^n$. Then for any $\varphi \in PSH(\Omega)$ and f holomorphic in $\Omega' := \Omega \cap H$ there exists a holomorphic extension F of f in Ω satisfying*

$$\int_\Omega |F|^2 e^{-\varphi} d\lambda \leq C \int_{\Omega'} |f|^2 e^{-\varphi} d\lambda',$$

where C is a constant depending only on n and the diameter of Ω.

The original proof from [98] was very complicated: it used abstract Kähler geometry and nontrivial Kähler identities. It was subsequently simplified by Siu [109] and Berndtsson [4]. The big breakthrough came recently with a very short proof by Chen [40] who was the first one to succeed in deducing the Ohsawa–Takegoshi theorem directly from Hörmander's estimate. In fact he proved even a slightly more general result, obtained earlier by McNeal and Varolin [91] with more complicated methods:

Theorem 3.2 *Assume that $\Omega \subset \mathbb{C}^{n-1} \times \Delta$ is pseudoconvex and let $H := \{z_n = 0\}$. Then for any $\varphi \in PSH(\Omega)$ and f holomorphic in $\Omega' := \Omega \cap H$ there exists a holomorphic extension F of f in Ω satisfying*

$$\int_\Omega \frac{|F|^2}{|z_n|^2 \log^2 |z_n|} e^{-\varphi} d\lambda \leq C \int_{\Omega'} |f|^2 e^{-\varphi} d\lambda',$$

where C is an absolute constant.

Note that Theorem 3.2 clearly implies Theorem 3.1 by iteration and since $|\zeta|^2 \log^2 |\zeta|$ is bounded in Δ. Theorem 3.2 will easily follow from Theorem 2.3 and the following completely elementary lemma:

Lemma 3.3 *For $\zeta \in \mathbb{C}$ with $|\zeta| \leq (2e)^{-1/2}$ and $\varepsilon > 0$ sufficiently small set*

$$\psi(\zeta) := -\log\left[-\log\left(|\zeta|^2 + \varepsilon^2\right) + \log\left(-\log\left(|\zeta|^2 + \varepsilon^2\right)\right)\right].$$

Then ψ is subharmonic in $\{|\zeta| < (2e)^{-1/2}\}$ and there exist constants C_1, C_2, C_3 such that

(i) $\left(1 - \dfrac{|\psi_\zeta|^2}{\psi_{\zeta\bar\zeta}}\right) e^\psi \geq \dfrac{1}{C_1 \log^2(|\zeta|^2 + \varepsilon^2)}$ on $\{|\zeta| \leq (2e)^{-1/2}\}$;

(ii) $\dfrac{|\psi_\zeta|^2}{\psi_{\zeta\bar\zeta}} \leq \dfrac{C_2}{-\log \varepsilon}$ on $\{|\zeta| \leq \varepsilon\}$;

(iii) $\dfrac{e^\psi}{|\zeta|^2 \psi_{\zeta\bar\zeta}} \leq C_3$ on $\{\varepsilon/2 \leq |\zeta| \leq \varepsilon\}$.

Proof Write $t = 2\log|\zeta|$ and let γ be such that $\psi = \gamma(t)$. That is

$$\gamma = -\log(-\delta + \log(-\delta)),$$

where $\delta = -\log(e^t + \varepsilon^2)$. We have $\psi_\zeta = \gamma'/\zeta$, $\psi_{\zeta\bar\zeta} = \gamma''/|\zeta|^2$ and thus

$$\frac{|\psi_\zeta|^2}{\psi_{\zeta\bar\zeta}} = \frac{(\gamma')^2}{\gamma''}.$$

We have to prove that

$$\left(1 - \frac{(\gamma')^2}{\gamma''}\right) \geq \frac{-\delta + \log(-\delta)}{C_1 \delta^2} \qquad \text{if } t \leq -\log(2e) \qquad (3.1)$$

$$\frac{(\gamma')^2}{\gamma''} \leq \frac{C_2}{-\log \varepsilon} \qquad \text{if } t \leq 2\log\varepsilon \qquad (3.2)$$

$$(-\delta + \log(-\delta))\gamma'' \geq \frac{1}{C_3} \qquad \text{if } 2\log(\varepsilon/2) \leq t \leq 2\log\varepsilon. \qquad (3.3)$$

We can compute that

$$\gamma' = \frac{1 - \delta^{-1}}{-\delta + \log(-\delta)}\delta'$$

and

$$\gamma'' \geq \frac{1 - \delta^{-1}}{-\delta + \log(-\delta)}\delta''.$$

Therefore we get (3.3) and since

$$\frac{(\gamma')^2}{\gamma''} \leq \frac{1 - \delta^{-1}}{-\delta + \log(-\delta)}\frac{(\delta')^2}{\delta''} = \frac{1 - \delta^{-1}}{-\delta + \log(-\delta)}e^t,$$

we also obtain (3.1) and (3.2). □

Proof of Theorem 3.2 It will be no loss of generality to prove the result in a slightly smaller disc than Δ, say the same as in Lemma 3.3. By approximation we may assume that Ω is bounded, smooth, strongly pseudoconvex, φ is smooth up to the boundary, and f is holomorphic in a neighborhood of $\overline{\Omega'}$. Let $\chi \in C^\infty(\mathbb{R})$ be such that $\chi(t) = 1$ for $t \le -2$, $\chi(t) = 0$ for $t \ge 0$, and $|\chi'| \le 1$. For $\varepsilon > 0$ sufficiently small the function fv, where

$$v = v_\varepsilon := \chi(2\log(|z_n|/\varepsilon)),$$

is defined in Ω. We will use Theorem 2.3 for

$$\alpha = \alpha_\varepsilon := \bar{\partial}(fv) = f\,\chi'(2\log(|z_n|/\varepsilon))\frac{d\bar{z}_n}{\bar{z}_n},$$

$\widetilde{\varphi} := \varphi + 2\log|z_n|$, and ψ as in Lemma 3.3. We will find $u = u_\varepsilon \in L^2_{loc}(\Omega)$ such that $\bar{\partial}u = \alpha$ (in fact u has to be continuous, since fv is) and

$$\int_\Omega (1 - |\bar{\partial}\psi|^2_{i\partial\bar{\partial}\psi})|u|^2 e^{\psi - \widetilde{\varphi}} d\lambda \le \frac{1 + \sqrt{a}}{1 - \sqrt{a}}\int_\Omega |\alpha|^2_{i\partial\bar{\partial}\psi} e^{\psi - \widetilde{\varphi}} d\lambda, \qquad (3.4)$$

where $a = -C_2/\log\varepsilon$ by Lemma 3.3ii. For a given ε the function

$$\left(1 - |\bar{\partial}\psi|^2_{i\partial\bar{\partial}\psi}\right)e^{\psi - \widetilde{\varphi}}$$

is not integrable near H, and thus by (3.4) $u = 0$ on Ω'. This means that $F_\varepsilon := fv - u$ is a holomorphic extension of f to Ω. (3.4) together with Lemma 3.3i also give

$$\int_\Omega \frac{|u|^2}{|z_n|^2 \log^2(|z_n|^2 + \varepsilon^2)} e^{-\varphi} d\lambda \le C_1 \frac{1 + \sqrt{a}}{1 - \sqrt{a}}\int_\Omega |\alpha|^2_{i\partial\bar{\partial}\psi} e^{\psi - \varphi} d\lambda'.$$

Using Lemma 3.3iii we will obtain

$$\limsup_{\varepsilon \to 0}\int_\Omega \frac{|F_\varepsilon|^2}{|z_n|^2 \log^2|z_n|}|e^{-\varphi}d\lambda \le C\int_{\Omega'}|f|^2 e^{-\varphi}d\lambda'$$

and it remains to apply the Banach–Alaouglu theorem.   □

4 Singularities of plurisubharmonic functions

We will start with the following recent result of Berndtsson [10] (proved by Favre and Jonsson [54] in dimension 2) confirming the *openness conjecture* of Demailly–Kollár [46].

Theorem 4.1 *For a plurisubharmonic function φ defined in a neighbourhood of $z_0 \in \mathbb{C}^n$ the set of those $p \in \mathbb{R}$ such that $e^{-p\varphi}$ is integrable near z_0 is an open interval of the form $(-\infty, p_0)$.*

The whole point is that the limit p_0 does not belong to this set. First of all it is easy to see that this holds for $n = 1$. Then φ can be written as a sum of a harmonic function and the potential

$$U^\mu(z) = \int_{\mathbb{C}} \log |\zeta - z| d\mu(\zeta),$$

where μ is a positive measure with compact support in $\mathbb{C}$ such that $\mu = \Delta\varphi/2\pi$ near z_0. We may thus assume that $\varphi = U^\mu$ and then one can then easily prove that $e^{-p\varphi}$ is integrable near z_0 if and only if $p\,\mu(\{z_0\}) < 2$.

The original proof of Theorem 3.1 from [10] was more complicated but Berndtsson [11] extracted the following simple one from the method of Guan–Zhou [58] who showed a more general *strong openness conjecture*, where instead of $e^{-p\varphi}$ one is interested in local integrability of $|f|^2 e^{-p\varphi}$ for some fixed holomorphic f. The proof of this was simplified by Hiep [64].

Proof of Theorem 4.1 We may assume that z_0 is the origin, φ is defined in a neighbourhood of $\bar{\Delta}^n$ and $\varphi \leq 0$. We first claim that if φ is not locally integrable near the origin then

$$\int_{\Delta^{n-1}} e^{-\varphi(\cdot, z_n)} d\lambda' \geq \frac{c_n}{|z_n|^2}, \quad |z_n| \leq 1/2, \tag{4.1}$$

where c_n is a positive constant depending only on n. For a fixed z_n we may assume that the left-hand side of (4.1) is finite. By the Ohsawa–Takegoshi theorem there exists a holomorphic F in Δ^n such that $F(\cdot, z_n) = 1$ in Δ^{n-1} and

$$\int_{\Delta^n} |F|^2 e^{-\varphi} d\lambda \leq C_1 \int_{\Delta^{n-1}} e^{-\varphi(\cdot, z_n)} d\lambda' < \infty. \tag{4.2}$$

It is elementary that

$$|F(0, \zeta)|^2 \leq C_2 \int_{\Delta^n} |F|^2 d\lambda \leq C_2 \int_{\Delta^n} |F|^2 e^{-\varphi} d\lambda, \quad |\zeta| \leq 1/2. \tag{4.3}$$

Since $e^{-\varphi}$ is not locally integrable near the origin, by (4.2) we have $F(0, 0) = 0$, and thus by (4.3) and the Schwarz lemma

$$|F(0, \zeta)|^2 \leq C_3 |\zeta|^2 \int_{\Delta^n} |F|^2 e^{-\varphi} d\lambda, \quad |\zeta| \leq 1/2.$$

For $\zeta = z_n$ using (4.2) and the fact that $F(0, z_n) = 1$ we get (4.1).

Now assume that the result is true for functions of $n - 1$ variables and suppose that

$$\int_{\Delta^n} e^{-p_0\varphi} d\lambda < \infty. \tag{4.4}$$

Since for $p > p_0$ we know that $e^{-p\varphi}$ is not locally integrable near the origin, by (4.1)

$$\int_{\Delta^{n-1}} e^{-p\varphi(\cdot, z_n)} d\lambda' \geq \frac{c_n}{|z_n|^2}, \quad |z_n| \leq 1/2. \tag{4.5}$$

From (4.4) it follows that for almost all $z_n \in \Delta$

$$\int_{\Delta^{n-1}} e^{-p_0\varphi(\cdot, z_n)} d\lambda' < \infty$$

and thus by the inductive assumption for p sufficiently close to p_0

$$\int_{\Delta^{n-1}} e^{-p\varphi(\cdot, z_n)} d\lambda' < \infty.$$

The Lebesgue dominated convergence theorem now implies that (4.5) also holds for $p = p_0$ which contradicts (4.4). □

It is quite remarkable that to prove a result on plurisubharmonic functions one has to use tools like holomorphic function and $\bar{\partial}$-equation.

For a plurisubharmonic φ defined in a neighborhood of z_0 its *Lelong number* at z_0 is defined by

$$\nu_\varphi(z_0) = \liminf_{z \to z_0} \frac{\varphi(z)}{\log|z - z_0|} = \lim_{r \to 0^+} \frac{\varphi^r(z_0)}{\log r},$$

where

$$\varphi^r(z) = \max_{|\zeta - z| \leq r} \varphi(\zeta). \tag{4.6}$$

(One can show that φ^r, defined in $\Omega_r := \{z \in \Omega : B(z, r) \subset \Omega\}$, is continuous, plurisubharmonic and decreases to φ as r decreases to 0.) In other words, $\nu_\varphi(z_0)$ is the maximal number $c \geq 0$ such that

$$\varphi(z) \leq c \log|z - z_0| + A$$

for some constant A and z in a neighbourhood of z_0. Lelong number measures the singularity of a plurisubharmonic function at a point.

The classical result on Lelong numbers is the following due to Siu [107]:

Theorem 4.2 *For any plurisubharmonic function φ and $c \in \mathbb{R}$ the superlevel set $\{\nu_\varphi \geq c\}$ is analytic.*

The original proof in [107] was very complicated. It was later simplified and generalized by Kiselman [75,77] (see also [66]) and Demailly [44]. It was Demailly [45] who found a surprisingly simple proof of the Siu theorem using the Ohsawa–Takegoshi theorem. It was done through the following approximation of plurisubharmonic functions:

Theorem 4.3 *Let φ be plurisubharmonic in a bounded pseudoconvex Ω in $\mathbb{C}^n$. For $m = 1, 2, \ldots$ define*

$$\varphi_m := \frac{1}{2m} \log \sup \left\{ |f|^2 : f \in \mathcal{O}(\Omega), \quad \int_\Omega |f|^2 e^{-2m\varphi} d\lambda \leq 1 \right\}.$$

Then there exist positive constants C_1 depending only on n and the diameter of Ω and C_2 depending only on n such that

$$\varphi - \frac{C_1}{m} \leq \varphi_m \leq \varphi^r + \frac{1}{m} \log \frac{C_2}{r^n} \quad in \quad \Omega_r \tag{4.7}$$

and

$$v_\varphi - \frac{n}{m} \leq v_{\varphi_m} \leq v_\varphi. \tag{4.8}$$

In particular, $\varphi_m \to \varphi$ pointwise and in L^1_{loc}.

Proof By the Ohsawa–Takegoshi theorem for every $z \in \Omega$ we can find $f \in \mathcal{O}(\Omega)$ such that

$$\int_\Omega |f|^2 e^{-2m\varphi} d\lambda \leq C |f(z)|^2 e^{-2m\varphi(z)} = 1.$$

This implies that

$$\varphi_m(z) \geq \frac{1}{2m} \log |f(z)|^2 = \varphi(z) - \frac{\log C}{2m}$$

and we obtain the first inequality in (4.7). The proof of the second one is completely elementary: $|f|^2$ is in particular subharmonic and thus for $r < \operatorname{dist}(z, \partial\Omega)$

$$|f(z)|^2 \leq \frac{1}{\lambda(B(z,r))} \int_{B(z,r)} |f|^2 d\lambda \leq \frac{n!}{\pi^n r^{2n}} e^{2m\varphi^r(z)} \int_\Omega |f|^2 e^{-2m\varphi} d\lambda$$

which gives the second inequality in (4.7).

Now (4.8) easily follows from (4.7): the first inequality in (4.7) implies that $v_{\varphi_m} \leq v_{\varphi - C_1/m} = v_\varphi$ and the second one gives

$$\varphi_m^r \leq \varphi^{2r} + \frac{1}{m} \log \frac{C_2}{r^n},$$

hence $v_\varphi - n/m \leq \varphi_{n/m}$. $\qquad\qquad\qquad\qquad\qquad\qquad\qquad\qquad\qquad\qquad\qquad\qquad\square$

Proof of Theorem 4.2 The result is local so we may assume that φ is defined in bounded pseudoconvex domain Ω. Then by (4.8)

$$\{v_\varphi \geq c\} = \bigcap_m \left\{ v_{\varphi_m} \geq c - \frac{n}{m} \right\}.$$

Let $\{\sigma_j\}$ be an orthonormal basis of $\mathcal{O}(\Omega) \cap L^2(\Omega, e^{-2m\varphi})$. Then

$$\varphi_m = \frac{1}{2m} \log \sum_j |\sigma_j|^2 \tag{4.9}$$

and one can show that

$$\left\{ v_{\varphi_m} \geq c - \frac{n}{m} \right\} = \bigcap_{\substack{|\alpha| < mc-n \\ j}} \{\partial^\alpha \sigma_j = 0\}$$

which finishes the proof.   □

It is interesting that the Ohsawa–Takegoshi theorem also gives the following sub-additivity of the Demailly approximation from [47]:

Theorem 4.4 *Under the assumptions of Theorem 4.3 there exists a positive constant C_3 depending only on n and the diameter of Ω such that*

$$(m_1 + m_2)\varphi_{m_1+m_2} \leq m_1\varphi_{m_1} + m_2\varphi_{m_2} + C_3. \tag{4.10}$$

Proof By the Ohsawa–Takegoshi theorem for every $f \in \mathcal{O}(\Omega)$ with

$$\int_\Omega |f|^2 e^{-2(m_1+m_2)\varphi} d\lambda \leq 1$$

there exists $F \in \mathcal{O}(\Omega \times \Omega)$ such that $F(z, z) = f(z)$ for $z \in \Omega$ and

$$\iint_{\Omega \times \Omega} |F(z, w)|^2 e^{-2m_1\varphi(z)-m_2\varphi(w)} d\lambda(z)d\lambda(w) \leq C. \tag{4.11}$$

Let $\{\sigma_j\}$ be an orthonormal basis in $\mathcal{O}(\Omega) \cap L^2(\Omega, e^{-2m_1\varphi})$ and $\{\sigma'_k\}$ an orthonormal basis in $\mathcal{O}(\Omega) \cap L^2(\Omega, e^{-2m_2\varphi})$, then $\{\sigma_j(z)\sigma'_k(w)\}$ is an orthonormal basis in $\mathcal{O}(\Omega \times \Omega) \cap L^2(\Omega \times \Omega, e^{-2m_1\varphi(z)-2m_2\varphi(w)})$. If

$$F(z, w) = \sum_{j,k} c_{jk}\sigma_j(z)\sigma'_k(w)$$

then by (4.11) $\sum_{j,k} |c_{jk}|^2 \leq C$ and thus by the Schwarz inequality and (4.9)

$$|f(z)|^2 = |F(z, z)|^2 \leq C \sum_j |\sigma_j(z)|^2 \sum_k |\sigma'_k(z)|^2 = Ce^{2m_1\varphi_{m_1}(z)}e^{2m_2\varphi_{m_2}(z)}.$$

This gives (4.10) with $C_3 = \log C/2$.   □

Theorem 4.4 gives monotonicity of a subsequence of φ_m. More precisely, for example the sequence $\varphi_{2^k} + C_3/2^{k+1}$ is decreasing. It was recently showed by Kim [74] that in general one cannot expect monotonicity of the entire sequence φ_m, even after adding a sequence of constants converging to 0.

5 Pluricomplex Green function and the complex Monge–Ampère operator

If Ω is an open subset of $\mathbb{C}^n$ then for $z, w \in \Omega$ the *pluricomplex Green function* is defined as

$$G_\Omega(z, w) = \sup\{u(z) \colon u \in \mathcal{B}(\Omega, w)\},$$

where $\mathcal{B}(\Omega, w)$ is the family of negative plurisubharmonic functions in Ω that have a logarithmic pole at w, that is

$$\mathcal{B}(\Omega, w) = \left\{u \in PSH^-(\Omega) \colon \limsup_{z \to w}(u(z) - \log|z - w|) < \infty\right\}.$$

One can show that for a given $w \in \Omega$ we either have $G_\Omega(\cdot, w) \in \mathcal{B}(\Omega, w)$ or $\mathcal{B}(\Omega, w) = \emptyset$. This general definition of the pluricomplex Green function was first given independently by Klimek [78] and Zakharyuta [115]. The fundamental properties were proved by Demailly [43].

One of the big differences between one and higher dimensional cases is that for $n \geq 2$ the Green function is usually not symmetric. The first example of this kind is due to Bedford and Demailly [3]. The following simple one was given by Klimek [79]: for $\Omega = \{|z_1 z_2| < 1\} \subset \mathbb{C}^2$ one can show that

$$G_\Omega(z, w) = \begin{cases} \log\left|\frac{z_1 z_2 - w_1 w_2}{1 - \bar{w}_1 \bar{w}_2 z_1 z_2}\right| & w \neq 0, \\ \frac{1}{2} \log|z_1 z_2| & w = 0. \end{cases}$$

In particular, $G_\Omega(z, 0) = \frac{1}{2} \log|z_1 z_2|$ but $G_\Omega(0, z) = \log|z_1 z_2|$. On the other hand, it follows from Lempert's theory [85] that G_Ω is symmetric for convex Ω.

The main tool when dealing with the pluricomplex Green function is Bedford–Taylor's theory of the complex Monge–Ampère operator [1,2]. It is convenient to consider the operators $d = \partial + \bar{\partial}$ and $d^c := i(\bar{\partial} - \partial)$, so that $dd^c = 2i\partial\bar{\partial}$. For smooth u we then have

$$(dd^c u)^n = dd^c u \wedge \cdots \wedge dd^c u = 4^n n! \det(\partial^2 u/\partial z_j \partial \bar{z}_k)\, d\lambda$$

and one would like to define $(dd^c u)^n$ as a positive regular measure for arbitrary plurisubharmonic u. This turned out to be impossible in general. First example was found by Shiffman and Taylor, see [108]. This was later simplified by Kiselman [76]: for $n \geq 2$ the function

$$u(z) = (-\log|z_1|)^{1/n}(|z_2|^2 + \cdots + |z_n|^2 - 1)$$

is plurisubharmonic near the origin, smooth away from $\{z_1 = 0\}$ but $(dd^c u)^n$ is not locally integrable near $\{z_1 = 0\}$.

Bedford and Taylor [2] proved however that it is possible to define $(dd^c u)^n$ for locally bounded plurisubharmonic u and Demailly [43] extended this to plurisubharmonic functions that are possibly unbounded on a compact subset. In both cases the operator $(dd^c)^n$ is continuous in the weak* topology of measures for monotone sequences. In fact, the domain of definition of the complex Monge–Ampère operator, defined as the maximal subclass of the class of plurisubharmonic functions where the operator can be defined as a positive measure in such a way that it is continuous for decreasing sequences, was characterized in [22] and [24]. In particular, for $n = 2$ these are precisely the plurisubharmonic functions which belong to the Sobolev space $W^{1,2}_{loc}$.

A plurisubharmonic function u in Ω is called *maximal* if for any other $v \in PSH(\Omega)$ such that $v \leq u$ in $\Omega \backslash K$ for some $K \Subset \Omega$ we have $v \leq u$ in Ω. For $n = 1$ these are precisely harmonic functions but they may be completely irregular in higher dimensions: for example if a plurisubharmonic function is independent of one of the variables then it is maximal. One of the main points of Bedford–Taylor's pluripotential theory [1,2] is that for locally bounded plurisubharmonic functions u we have

$$u \text{ is maximal } \Leftrightarrow (dd^c u)^n = 0. \tag{5.1}$$

The same characterization remains true for functions from the domain of definition of $(dd^c)^n$ (see [22]) but there are maximal plurisubharmonic functions which do not belong to the domain of definition, for example $\log |z_1|$ in $\mathbb{C}^n$ for $n \geq 2$. It remains an open problem whether maximality is a local property in general. By the above characterization as a solution to the homogeneous complex Monge–Ampère equation, it is true for locally bounded plurisubharmonic functions, or more generally functions from the domain of definition.

One can show that

$$G_{B(w,R)}(z, w) = \log \frac{|z - w|}{R}$$

and thus if $B(w, r) \subset \Omega \subset B(w, R)$ then

$$\log \frac{|z - w|}{R} \leq G_\Omega(z, w) \leq \log \frac{|z - w|}{r}.$$

Therefore, if Ω is bounded then for $w \in \Omega$ the function $G_\Omega(\cdot, w)$ is plurisubharmonic and locally bounded in $\Omega \backslash \{w\}$. We can then define the Monge–Ampère operator and Demailly [43] proved that

$$(dd^c G_\Omega(\cdot, w))^n = (2\pi)^n \delta_w \tag{5.2}$$

(see also [20]).

A domain Ω in $\mathbb{C}^n$ is called *hyperconvex* if it admits a negative plurisubharmonic exhaustion function, that is there exists $u \in PSH^-(\Omega)$ such that $\{u < t\} \Subset \Omega$ for $t < 0$. For $n = 1$ this equivalent to Ω being regular with respect to classical potential theory. In general, Kerzman and Rosay [73] proved that hyperconvexity is a local property of the boundary and Demailly [43] showed that pseudoconvex domains with Lipschitz boundary are hyperconvex. It is an open problem whether pseudoconvex domains with continuous boundary have to be hyperconvex.

Demailly [43] showed that if Ω is bounded and hyperconvex then G_Ω is continuous on $\bar{\Omega} \times \Omega$ away from the diagonal of Ω, where we extend the definition of G_Ω to vanish on $\partial\Omega \times \Omega$ (see also [19] for a slightly different proof). It is an open problem whether in this case G_Ω is continuous on $\bar{\Omega} \times \bar{\Omega}$ away from the diagonal of $\bar{\Omega}$. Equivalently, we ask whether for bounded hyperconvex Ω if $w_j \in \Omega$ is a sequence of poles converging to $\partial\Omega$ then $G_\Omega(\cdot, w_j)$ converge locally uniformly to 0. We have the following weaker result from [30]:

Proposition 5.1 *Assume that Ω is bounded and hyperconvex. Then for any $p < \infty$*

$$\lim_{w \to \partial\Omega} ||G_\Omega(\cdot, w)||_{L^p(\Omega)} = 0.$$

Proof By [15] there exists unique $u \in PSH(\Omega) \cap C(\bar{\Omega})$ such that $u = 0$ on $\partial\Omega$ and $(dd^c u)^n = d\lambda$. Write $G_w = G_\Omega(\cdot, w)$. Integrating by parts as in [14] we will get using (5.2)

$$\int_\Omega |G_w|^n d\lambda = \int_\Omega |G_w|^n (dd^c u)^n \leq n! ||u||_{L^\infty(\Omega)}^{n-1} \int_\Omega |u| (dd^c G_w)^n \leq C|u(w)|,$$

where C depends only on n and the volume of Ω. This gives the result for $p = n$ and for other p it follows easily from it. $\qquad\square$

The conjecture on locally uniform convergence of the Green function for poles converging to the boundary was confirmed by Herbort [63] for pseudoconvex domains with C^2 boundary (see also [23] for a slightly simplified proof). As in Proposition 5.1, the inequality for the complex Monge–Ampère operator from [14] is one of the tools. In fact, the only additional regularity of Ω used to prove this result is an existence of $u \in PSH(\Omega)$ such that

$$\frac{1}{A} \delta_\Omega(z)^a \leq |u(z)| \leq A \delta_\Omega(z)^b \tag{5.3}$$

for some positive constants A, a, b, where δ_Ω is the Euclidean distance to the boundary. For domains with C^2 boundary this is guaranteed by a theorem of Diederich and Fornæss [48], even with $a = b$. Since Harrington [61] generalized this Diederich–Fornæss result to pseudoconvex domains with Lipschitz boundary, the conjecture also holds in this case.

Further regularity of the pluricomplex Green function was established in [56] and [18] (see also [19]): if Ω is $C^{2,1}$-smooth and strongly pseudoconvex then for a fixed $w \in \Omega$ we have $G_\Omega(\cdot, w) \in C^{1,1}(\bar{\Omega}\backslash\{w\})$. This is the highest regularity we can expect, Bedford and Demailly showed that $G_\Omega(\cdot, w)$ does not have to be C^2-smooth up to the boundary even if Ω is C^∞-smooth and strongly pseudoconvex. Lempert [85] proved that $G_\Omega(\cdot, w) \in C^\infty(\bar{\Omega}\backslash\{w\})$ if Ω is C^∞-smooth and strongly convex.

The following result from [15] was used in the proof of Proposition 5.1: for any bounded hyperconvex Ω in $\mathbb{C}^n$ and nonnegative $F \in C(\bar{\Omega})$ there exists unique solution to the following Dirichlet problem:

$$\begin{cases} u \in PSH(\Omega) \cap C(\bar{\Omega}) \\ (dd^c u)^n = F\, d\lambda \\ u = 0 \quad \text{on } \partial\Omega \end{cases} \qquad (5.4)$$

It is an open problem whether the following interior regularity holds here: does $F \in C^\infty(\bar{\Omega})$ imply $u \in C^\infty(\Omega)$ (without any additional assumption on the regularity of Ω)? Of course when Ω is smooth and strongly pseudoconvex then it follows from the seminal work of Krylov [83] and Caffarelli et al. [37] that $u \in C^\infty(\bar{\Omega})$. In general however we cannot expect u to be smooth up to the boundary. The only case so far of a non-smooth domain where this problem was solved is a polydisk, see [16]. The main tool was transitivity of the group of holomorphic automorphisms used to show interior $C^{1,1}$-regularity, as in the classical result of Bedford and Taylor [1] for a ball. The corresponding result for the real Monge–Ampère equation in arbitrary bounded convex domain in $\mathbb{R}^n$ holds by the famous interior estimate of Pogorelov [99].

In Sect. 7 we will need the following product property of the pluricomplex Green function proved by Jarnicki and Pflug [69]:

Theorem 5.2 *Assume that $\Omega_j \subset \mathbb{C}^{n_j}$, $j = 1, 2$, are pseudoconvex. Then*

$$G_{\Omega_1 \times \Omega_2}\left(\left(z^1, z^2\right), \left(w^1, w^2\right)\right) = \max\left\{ G_{\Omega_1}\left(z^1, w^1\right), G_{\Omega_2}\left(z^2, w^2\right) \right\}. \quad (5.5)$$

Proof Directly from the definition we have $\geq$. To show $\leq$ we may assume that Ω_j are bounded hyperconvex. Then it is enough to show that for fixed $w^j \in \Omega_j$ the right-hand side od (5.5), as a function of (z^1, z^2), is maximal in $\Omega_1 \times \Omega_2 \setminus \{(w^1, w^2)\}$. By (5.1) we have to prove that it solves the homogeneous complex Monge–Ampère equation. This follows from the following result of Zeriahi [116]:

$$(dd^c u_j)^{n_j} = 0 \implies \left(dd^c \max\left\{ u_1\left(z^1\right), u_2\left(z^2\right) \right\} \right)^{n_1 + n_2} = 0$$

which can be easily deduced from the following formula originally proved in [17]:

Theorem 5.3 *Let u, v be locally bounded plurisubharmonic functions defined on an open subset of $\mathbb{C}^n$ and $2 \le p \le n$. Then*

$$(dd^c \max\{u, v\})^p = dd^c \max\{u, v\} \wedge \sum_{k=0}^{p-1} (dd^c u)^k \wedge (dd^c v)^{p-1-k}$$

$$- \sum_{k=1}^{p-1} (dd^c u)^k \wedge (dd^c v)^{p-k}.$$

Proof By approximation we may assume that u, v are smooth. A simple inductive argument reduces the proof to the case $p = 2$. Set $w := \max\{u, v\}$ and, for $\varepsilon > 0$, $w_\varepsilon := \max\{u + \varepsilon, v\}$. In an open set $\{u + \varepsilon > v\}$ we have $w_\varepsilon - u = \varepsilon$, whereas $w - v = 0$ in $\{u < v\}$. It follows that for every $\varepsilon > 0$ one has $dd^c(w_\varepsilon - u) \wedge dd^c(w - v) = 0$ and taking the limit we conclude that $dd^c(w - u) \wedge dd^c(w - v) = 0$. $\qquad\square$

Edigarian [53] showed Theorem 5.2 without assuming pseudoconvexity. His proof however is much more complicated, it uses Poletsky's theory of analytic disks [100].

6 Bergman completeness

For a domain Ω in $\mathbb{C}^n$ we set $A^2(\Omega) := \mathcal{O}(\Omega) \cap L^2(\Omega)$. It is a closed subspace of $L^2(\Omega)$ and thus a Hilbert space. It is conjectured that when Ω is pseudoconvex then either $A^2(\Omega) = \{0\}$ or $A^2(\Omega)$ is infinitely dimensional. Wiegerinck [114] showed this for $n = 1$ and found non-pseudoconvex Ω with $A^2(\Omega)$ of arbitrary dimension.

For $w \in \Omega$ the functional

$$A^2(\Omega) \ni f \longmapsto f(w) \in \mathbb{C}$$

is bounded and thus $f(w) = \langle f, K_w \rangle$ for some $K_w \in A^2(\Omega)$ and all f. The Bergman kernel is characterized by the reproducing formula

$$f(w) = \int_\Omega f(z) \overline{K_\Omega(z, w)} d\lambda(z), \quad f \in A^2(\Omega), \ w \in \Omega.$$

Applying this for $f = K_\Omega(\cdot, z)$ we see that K_Ω is antisymmetric:

$$K_\Omega(w, z) = \overline{K_\Omega(z, w)}$$

and

$$K_\Omega(z, z) = ||K_\Omega(\cdot, z)||^2 = \sup\{|f(z)|^2 : f \in A^2(\Omega), \ ||f|| \le 1\}, \qquad (6.1)$$

where $|| \cdot ||$ is the L^2-norm in Ω. By Hartogs' theorem on separate holomorphic functions K_Ω is smooth on $\Omega \times \Omega$. If $\{\sigma_j\}$ is an orthonormal system in $A^2(\Omega)$ then

$$K_\Omega(z, w) = \sum_j \sigma_j(z) \overline{\sigma_j(w)}$$

and on the diagonal

$$K_\Omega(z, z) = \sum_j |\sigma_j(z)|^2. \qquad (6.2)$$

For other basic properties of K_Ω we refer to [70].

For a big class of domains, e.g. bounded ones, on the diagonal we have $K_\Omega > 0$ and thus $\log K_\Omega(z, z)$ is a smooth plurisubharmonic function in Ω. If it is also strongly plurisubharmonic then we say that Ω *admits* the Bergman metric and the Kähler metric defined by the potential $\log K_\Omega(z, z)$ is called the *Bergman metric* of Ω. One can show that the Levi form is given by the following extremal formula

$$\sum_{p,q=1}^{n} \frac{\partial^2 (\log K_\Omega(z,z))}{\partial z_p \partial \bar{z}_q} X_p \bar{X}_q$$

$$= \frac{1}{K_\Omega(z,z)} \sup \left\{ |D_X f(z)|^2 : f \in A^2(\Omega), \ f(z) = 0, \ \|f\| \le 1 \right\},$$

where $D_X = \sum_p X_p \partial/\partial z_p$, and it follows easily that for example all bounded domains admit the Bergman metric.

If Ω is complete with respect to the geodesic distance defined by the Bergman metric then we say that Ω is *Bergman complete*. The main tool in studying Bergman completeness is the following embedding of Kobayashi [80]:

$$\kappa : \Omega \ni z \longmapsto [K_\Omega(\cdot, z)] \in \mathbb{P}(A^2(\Omega)).$$

One can easily show that if Ω admits the Bergman metric then κ is an immersion and if Ω is bounded then it is an embedding. The main point is that the pull-back of the Fubini–Study metric on the (infinitely dimensional) projective space $\mathbb{P}(A^2(\Omega))$ by κ is precisely the Bergman metric of Ω. This is sometimes called Kobayashi's alternative definition of the Bergman metric. An immediate consequence of this is that κ is distance decreasing which means that

$$\operatorname{dist}_\Omega^B(z,w) \ge \arccos \frac{|K_\Omega(z,w)|}{\sqrt{K_\Omega(z,z)K_\Omega(w,w)}}, \tag{6.3}$$

where $\operatorname{dist}_\Omega^B$ is the distance defined by the Bergman metric. In particular,

$$K_\Omega(z,w) = 0 \ \Rightarrow \ \operatorname{dist}_\Omega^B(z,w) \ge \frac{\pi}{2}$$

and Dinew [50] showed that $\pi/2$ is an optimal constant here.

We have the following criterion of Kobayashi [80] for Bergman completeness:

Theorem 6.1 *Assume that Ω admits the Bergman metric and is such that for any sequence $z_j \in \Omega$ without accumulation point in Ω we have*

$$\lim_{j \to \infty} \frac{|f(z_j)|^2}{K_\Omega(z_j, z_j)} = 0, \quad f \in A^2(\Omega). \tag{6.4}$$

Then Ω is Bergman complete.

Proof Assume that $z_j \in \Omega$ is a Cauchy sequence with respect to $\operatorname{dist}_\Omega^B$. If it has an accumulation point in Ω then it has a limit, since locally the Bergman metric is equivalent to the Euclidean metric. We may thus assume that it has no accumulation point in Ω. Since κ is distance decreasing, it follows that $\kappa(z_j)$ is a Cauchy sequence in $\mathbb{P}(A^2(\Omega))$ and thus has a limit there, say $[f]$ for some $f \in A^2(\Omega)$, $f \not\equiv 0$. This means that there exist $a_j \in \mathbb{C}$ such that

$$a_j K_\Omega(\cdot, z_j) \to f.$$

This gives $|a_j|\sqrt{K_\Omega(z_j, z_j)} \to ||f||$ and $|a_j|\,|f(z_j)| \to ||f||^2$ which imply that

$$\frac{|f(z_j)|^2}{K_\Omega(z_j, z_j)} \to ||f||^2,$$

a contradiction.   □

We say that a bounded Ω is *Bergman exhaustive* if

$$\lim_{z \to \partial\Omega} K_\Omega(z, z) = \infty.$$

Note that bounded domains satisfying (6.4) must be Bergman exhaustive, simply take $f \equiv 1$. The Hartogs triangle

$$\{z \in \mathbb{C}^2 : |z_2| < |z_1| < 1\}$$

is an example of a domain which is Bergman exhaustive but not Bergman complete. This can be shown using the fact that the Hartogs triangle is biholomorphic to $\Delta \times \Delta_*$. This example also shows that Bergman exhaustiveness is not a biholomorphic invariant, contrary to Bergman completeness. On the other hand, Chen [39] proved that for $n = 1$ Bergman exhaustiveness does imply Bergman completeness.

Zwonek [118] showed that the converse to Theorem 6.1 does not hold: he gave an example of a bounded domain in $\mathbb{C}$ which is Bergman complete but not Bergman exhaustive. This example was simplified by Jucha [72]: he showed that

$$\Omega := \Delta_* \setminus \left(\bigcup_{k=1}^\infty \bar{\Delta}(2^{-k}, r_k) \right),$$

where $r_k > 0$ are such that $\bar{\Delta}(2^{-k}, r_k) \cap \bar{\Delta}(2^{-l}, r_l) = \emptyset$ for $k \neq l$, is Bergman complete if and only if

$$\sum_{k=1}^\infty \frac{2^k}{\sqrt{-\log r_k}} = \infty$$

and Bergman exhaustive if and only if

$$\sum_{k=1}^\infty \frac{4^k}{-\log r_k} = \infty.$$

Therefore, if for example $r_k = e^{-k^2 4^k}$ then Ω is Bergman complete but not Bergman exhaustive.

The proof of Theorem 6.1 really shows something slightly stronger: instead of (6.4) it is enough to assume that

$$\varlimsup_{j \to \infty} \frac{|f(z_j)|^2}{K_\Omega(z_j, z_j)} < ||f||^2, \quad f \in A^2(\Omega), \ f \not\equiv 0.$$

It is not known if this condition is equivalent to Bergman completeness or not. Another open problem is whether Bergman exhaustiveness is a biholomorphically invariant notion for $n = 1$. In view of Chen's result, an example showing that it is not would be another one showing that (6.4) is not equivalent to Bergman completeness.

It turns out that pluripotential theory gives a lot of examples of Bergman complete domains. The main result is due to Chen [38] in dimension one and independently to Herbort [62] and Pflug et al. [30] in arbitrary dimension:

Theorem 6.2 *Bounded hyperconvex domains are Bergman complete.*

We will prove this using the following estimate of Herbort [62]:

Theorem 6.3 *Assume that Ω is pseudoconvex. Then for every $f \in A^2(\Omega)$ and $w \in \Omega$ one has*

$$\frac{|f(w)|^2}{K_\Omega(w, w)} \le c_n \int_{\{G_\Omega(\cdot, w) < -1\}} |f|^2 d\lambda. \tag{6.5}$$

Proof Approximating Ω from inside we may assume that it is bounded and hyperconvex. We will use Theorem 2.2 with

$$\varphi = 2nG, \quad \psi = -\log(-G),$$

and

$$\alpha = \bar{\partial}(f \chi \circ G) = f \chi' \circ G \, \bar{\partial} G,$$

where $G = G_\Omega(\cdot, w)$ and $\chi \in C^\infty((-\infty, 0))$ is such that $\chi(t) = 0$ for $t \ge -1/2$ and $\chi(t) = -1$ for $t \le -2$. We have

$$i\bar{\alpha} \wedge \alpha \le |f|^2 G^2 (\chi' \circ G)^2 i \partial \bar{\partial} \psi$$

and thus by Theorem 2.2 there exists $u \in L^2_{loc}(\Omega)$ (in fact it has to be continuous) such that $\bar{\partial} u = \alpha$ and

$$\int_\Omega |u|^2 d\lambda \le \int_\Omega |u|^2 e^{-\varphi} d\lambda \le C \int_\Omega |f|^2 G^2 (\chi' \circ G)^2 e^{-2nG} d\lambda. \tag{6.6}$$

Since $e^{-\varphi}$ is not locally integrable near w, it follows that for $F := f\chi \circ G - u$ is holomorphic in Ω, $F(w) = f(w)$ and

$$\int_\Omega |F|^2 \le c_n \int_{\{G < -1\}} |f|^2 d\lambda.$$

$\square$

Proof of Theorem 6.2 By Proposition 5.1

$$\lim_{w \to \partial \Omega} \lambda(\{G_\Omega(\cdot, w) < -1\}) = 0$$

and thus by Theorem 6.3

$$\lim_{w \to \partial \Omega} \frac{|f(w)|^2}{K_\Omega(w, w)} = 0.$$

The result now follows from Kobayashi's criterion Theorem 6.1.   □

Taking $f \equiv 1$ in Herbort's estimate (6.5) we get

$$K_\Omega(w, w) \geq \frac{1}{c_n \lambda(\{G_\Omega(\cdot, w) < -1\})}. \tag{6.7}$$

The proof of Proposition 5.1 now gives for bounded hyperconvex domains

$$K_\Omega(w, w) \geq \frac{1}{C(n, \lambda(\Omega))|u(w)|}, $$

where u is the solution to (5.4) with $F \equiv 1$. This is an interesting lower bound for the Bergman kernel in terms of a solution to the complex Monge–Ampère equation and is in fact a quantitative version of the following result of Ohsawa [96]:

Theorem 6.4 *Bounded hyperconvex domains are Bergman exhaustive.*

It turns out that getting optimal constant in Herbort's estimate (6.5) and especially in (6.7) can be extremely useful. Herbort originally obtained the constant

$$c_n = 1 + 4e^{4n+3+R^2},$$

so it depended in addition on the diameter R of Ω. If we look at the proof of Theorem 6.3 closer and choose χ a bit more carefully then we can improve the constant obtained there considerably. Take $\chi \in C^{0,1}((-\infty, 0))$ such that $\chi(t) = 0$ for $t \geq -1$ and for $t < -1$ choose it in such a way that $t\chi'(t)e^{-nt} = -1$, that is

$$\chi(t) = \begin{cases} 0 & t \geq -1 \\ \int_1^{-t} \frac{ds}{se^{ns}} & t < -1 \end{cases}. \tag{6.8}$$

Then $F(w) = \chi(-\infty)f(w)$ and as in [23] we will get

$$c_n = \left(1 + \frac{C}{\text{Ei}(n)}\right)^2, \tag{6.9}$$

where

$$\mathrm{Ei}\,(a) = \int_a^\infty \frac{ds}{se^s}$$

and C is the constant from Theorem 2.2 (we know that $C = 4$ is optimal there). We will determine the optimal c_n in Sect. 7.

Bergman completeness of a bounded domain is equivalent to the fact that $\mathrm{dist}\,_\Omega^B(z, w) \to \infty$ as $z \to \partial\Omega$ and w is fixed. Theorem 6.2 does not give any quantitative version of this, even in terms of pluripotential theory. Diederich and Ohsawa [49] showed a lower bound for the Bergman distance for bounded pseudoconvex domains with C^2 boundary implying in particular completeness, this was later improved in [23]:

$$\mathrm{dist}\,_\Omega^B(z, w) \geq \frac{-\log \delta_\Omega(z)}{C \log(-\log \delta_\Omega(z))}, \tag{6.10}$$

where C is a positive constant depending only on Ω and w. The proof used the following estimate from [23] for the Bergman distance in terms of pluripotential theory:

Theorem 6.5 *Let Ω be pseudoconvex in $\mathbb{C}^n$ and assume that $z, w \in \Omega$ are such that*

$$\{G_\Omega(\cdot, z) < -1\} \cap \{G_\Omega(\cdot, w) < -1\} = \emptyset. \tag{6.11}$$

Then

$$\cdot\frac{|K_\Omega(z, w)|}{\sqrt{K_\Omega(z, z)K_\Omega(w, w)}} \leq \frac{1}{\sqrt{1 + a_n^2}}, \tag{6.12}$$

where

$$a_n = \left(1 + \frac{2e^n}{\mathrm{Ei}\,(n)}\right)^{-1},$$

and

$$\mathrm{dist}\,_\Omega^B(z, w) \geq \arctan a_n. \tag{6.13}$$

Proof First note that (6.13) follows directly from (6.12) and (6.3). The proof of (6.12) will be similar to that of Theorem 6.3. We may assume that Ω is bounded and hyperconvex. We will use Theorem 2.2 with

$$\varphi = 2n(G_z + G_w), \quad \psi = -\log(-G_z),$$

where $G_z = G_\Omega(\cdot, z)$. Set

$$f := \frac{K_\Omega(\cdot, w)}{\sqrt{K_\Omega(w, w)}} \in A^2(\Omega),$$

so that $||f|| = 1$, and

$$\alpha := \bar{\partial}(f\,\chi \circ G_z) = f\,\chi' \circ G_z\,\bar{\partial}G_z,$$

where χ is given by (6.8). We can find continuous u in Ω solving $\bar{\partial}u = \alpha$ and such that

$$\int_\Omega |u|^2 d\lambda \leq \int_\Omega |u|^2 e^{-\varphi} d\lambda \leq 4 \int_\Omega |f|^2 G_z^2 (\chi' \circ G_z)^2 e^{-2n(G_z+G_w)} d\lambda$$

$$\leq 4e^{2n} \int_{\{G_z < -1\}} |f|^2 d\lambda,$$

where the last inequality follows from (6.11). We thus get $||u|| \leq 2e^n$ (because $||f|| = 1$) and, since $e^{-\varphi}$ is not locally integrable near both z and w, that $u(z) = u(w) = 0$. The function $F = f\chi \circ G_z - u$ is thus holomorphic and such that $F(z) = \mathrm{Ei}\,(n) f(z)$, $F(w) = 0$ (the latter by (6.11) again). We also have $||F|| \leq \mathrm{Ei}\,(n) + 2e^n$.

By the definition of f

$$\langle F, f\rangle = \frac{F(w)}{\sqrt{K_\Omega(w,w)}} = 0.$$

Therefore by (6.2)

$$K_\Omega(z,z) \geq |f(z)|^2 + \frac{|F(z)|^2}{||F||^2} \geq |f(z)|^2 \left(1 + a_n^2\right),$$

and (6.12) follows.   □

Theorem 6.5 reduced the proof of (6.10) in [23] to right estimates for the pluri-complex Green function, as in [63]. Since the only information really needed is (5.3) with $a = b$, by [61] the estimate (6.10) also holds for pseudoconvex domains with Lipschitz boundary. It is an open problem whether (6.10) can be improved to

$$\mathrm{dist}_\Omega^B(z,w) \geq \frac{1}{C}(-\log \delta_\Omega(z)),$$

which would be optimal. This estimate is known to hold for smooth strongly pseudo-convex domains and also for convex ones (without any regularity assumption, see [23]).

Lu Qi-Keng [87] showed that if the Bergman metric has constant sectional curvature then it is biholmorphic to a ball. A conjecture of Cheng asserts that this assumption can be weakened for smooth strongly pseudoconvex domains. It states that such a domain is biholomorphic to a ball if and only if its Bergman metric is Kähler–Einstein, that is its Ricci curvature is proportional to the metric. For $n = 1$ it follows from [87] and for $n = 2$ it was shown by Nemirovskii and Shafikov [94]. It remains open in higher dimensions.

7 Suita conjecture

Let D be a domain in $\mathbb{C}$ admitting the Green function which means exactly that the complement of D is not polar. For $z \in D$ set

$$c_D(z) := \exp \lim_{\zeta \to z} (G_D(\zeta, z) - \log |\zeta - z|).$$

It is in fact the logarithmic capacity of the complement of D with respect to z and the function under the exponent is called the *Robin function* for G_D. The function c_D is not biholomorphically invariant but one can easily check that the metric $c_D|dz|$ does not depend on a local holomorphic change of variables and thus is an invariant metric even for Riemann surfaces. It is called the *Suita metric* of D. Its curvature is given by

$$Curv_{c_D|dz|} = -\frac{\partial^2 (\log c_D)/\partial z \partial \bar{z}}{c_D^2}.$$

Suita [110] conjectured that

$$Curv_{c_D|dz|} \leq -1. \tag{7.1}$$

It is easy to see that we have equality for a disk and thus for simply connected domains. Using elliptic functions Suita showed that one has strict inequality in (7.1) if D is an annulus, and thus also any regular doubly connected domain. In fact, for $D = \{e^{-5} < |z| < 1\}$ the graph of $Curv_{c_D|dz|}$ as a function of $\log |z|$ looks as follows:[1]

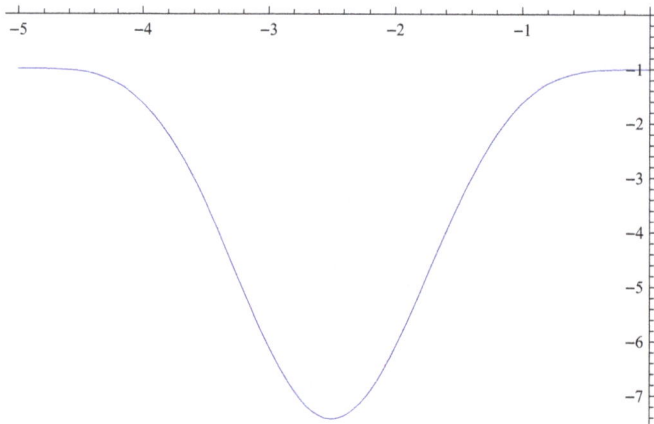

By approximation it is enough to verify (7.1) for bounded smooth D and then one can show that we have equality in (7.1) on the boundary. Therefore the Suita conjecture essentially asks whether the curvature of the Suita metric satisfies the maximum principle. This is in fact a rather rare situation for invariant metrics in

---

[1] Figures were obtained using *Mathematica*.

complex analysis, for example it is not satisfied for the Bergman metric. For the same annulus as before we will then have the following picture:

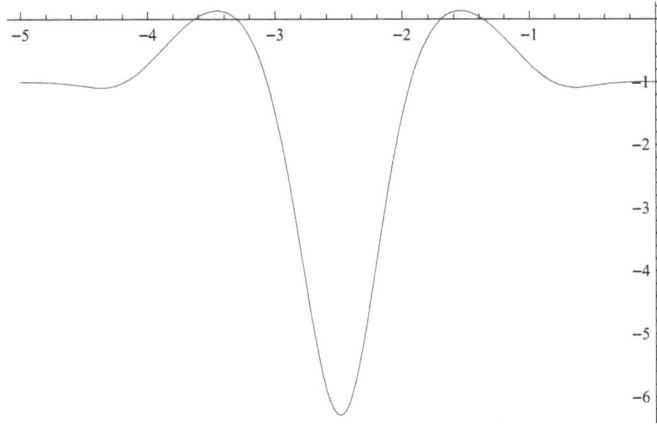

See [51] and [119] for specific results on the curvature of the Bergman metric on an annulus.

Surprisingly, it turned out that only the methods of several complex variables have given any real progress in this one-dimensional problem. It was the breakthrough of Ohsawa [97] who noticed that it is really an extension problem closely related to the Ohsawa–Takegoshi theorem. It was proved already by Suita [110] that

$$\frac{\partial^2 (\log c_D(z))}{\partial z \partial \bar{z}} = \pi K_D(z, z),$$

this in fact follows easily from the Schiffer formula

$$K_D(z, w) = \frac{2}{\pi} \frac{\partial^2 G_D(z, w)}{\partial z \partial \bar{w}}, \quad z \neq w,$$

and therefore (7.1) is equivalent to

$$c_D(z)^2 \leq \pi K_D(z, z). \tag{7.2}$$

But this is in turn equivalent to the following extension problem: for a given $z \in D$ find $f \in \mathcal{O}(D)$ such that $f(z) = 1$ and

$$\int_D |f(z)|^2 d\lambda \leq \frac{\pi}{c_D(z)^2}.$$

Ohsawa [97], using the same methods as in the original proof of the Ohsawa–Takegoshi theorem, proved the estimate

$$c_D(z)^2 \leq C K_D(z, z).$$

for some large absolute constant C. It was later improved in [25] and [59].

The optimal constant was eventually obtained in [27] where the following version of the Ohsawa–Takegoshi theorem also with optimal constant was proved:

Theorem 7.1 *Assume that D is a domain in $\mathbb{C}$ containing the origin. Let $\Omega \subset \mathbb{C}^{n-1} \times D$ be pseudoconvex, $\varphi \in PSH(\Omega)$, and set $\Omega' := \Omega \cap \{z_n = 0\}$. Then for any $f \in \mathcal{O}(\Omega')$ there exists $F \in \mathcal{O}(\Omega)$ such that $F = f$ on Ω' and*

$$\int_\Omega |F|^2 e^{-\varphi} d\lambda \leq \frac{\pi}{c_D(0)^2} \int_{\Omega'} |f|^2 e^{-\varphi} d\lambda'.$$

For $n = 1$ we obtain the Suita conjecture (7.2).

The proof of Theorem 7.1 was similar to that of Theorem 3.2 but Theorem 2.4 was used instead of Theorem 2.3 and the weights were chosen more carefully. Theorem 2.4 was used in [27] with weights of the form

$$\widetilde{\varphi} = \varphi + 2G + \eta(-2G), \quad \psi = \gamma(-2G),$$

where $G = G_D(\cdot, 0)$. It was rather straightforward, although technical, how to define $\eta(t)$ and $\gamma(t)$ for $t \geq -2\log \varepsilon$ (that is $\widetilde{\varphi}$ and ψ near $\{z_n = 0\}$). The main problem was to construct h, g on $(0, \infty)$ behaving like $-\log t$ near ∞ and such that

$$\left(1 - \frac{(g')^2}{h''}\right) e^{2g-h+t} \geq 1. \tag{7.3}$$

Eventually it turned out that solutions can be written explicitly:

$$h(t) := -\log(t + e^{-t} - 1)$$
$$g(t) := -\log(t + e^{-t} - 1) + \log(1 - e^{-t})$$

and we even have equality in (7.3). In fact, when a similar method was used earlier in [26] but with Theorem 2.3 instead of 2.4, it lead to an ODE with only one unknown:

$$\left(1 - \frac{(g')^2}{g''}\right) e^{g+t} \geq 1$$

and the best constant one can get this way is 1.95388..., the same as the one obtained earlier in [59].

After [27], Guan and Zhou [57] proved various generalizations of Theorem 7.1 but used essentially the same ODE with two unknowns as (7.3) and got essentially the same solutions. They also characterized precisely the case when there is equality in (7.2) answering a more precise question posed by Suita [110]:

Theorem 7.2 *Let M be a Riemann surface admitting the Green function (which is equivalent to the fact that there exists a bounded nonconstant subharmonic function on M). Then (7.2) holds and if we have equality for some $z \in M$ then M is biholomorphic to $\Delta \backslash F$ where F is a closed polar subset of Δ.*

Another approach to the Suita conjecture was presented in [28]. The idea was to obtain optimal constants in (6.7) for arbitrary sublevel sets. It turned out that the constant obtained already can be improved to the optimal one quite easily using the tensor power trick. The following general lower bound for the Bergman kernel on the diagonal was obtained:

Theorem 7.3 *Let Ω be pseudoconvex, $w \in \Omega$ and $t \leq 0$. Then*

$$K_\Omega(w, w) \geq \frac{1}{e^{-2nt}\lambda(\{G_\Omega(\cdot, w) < t\})}. \tag{7.4}$$

Proof Repeating the argument of the proof of Theorem 6.3 with the constant given by (6.9) for $f \equiv 1$ and arbitrary t we will obtain

$$K_\Omega(w, w) \geq \frac{c(n, t)}{\lambda(\{G_\Omega(\cdot, w) < t\})}, \tag{7.5}$$

where

$$c(n, t) = \left(1 + \frac{C}{\mathrm{Ei}\,(nt)}\right)^2$$

and C is the constant from Theorem 2.2. We now use the tensor power trick: for a positive integer m take $\widetilde{\Omega} = \Omega^m \subset \mathbb{C}^{nm}$ and $\widetilde{w} = (w, \ldots, w)$. Then by the product properties for the Bergman kernel (see e.g. [70]) and for the pluricomplex Green function, Theorem 5.2, we have

$$K_{\widetilde{\Omega}}(\widetilde{w}, \widetilde{w}) = (K_\Omega(w))^m, \quad \{G_{\widetilde{\Omega}}(\cdot, \widetilde{w}) < t\} = \{G_\Omega(\cdot, w) < t\}^m,$$

and thus by (7.5)

$$K_\Omega(w, w) \geq \frac{c(nm, t)^{1/m}}{\lambda(\{G_\Omega(\cdot, w) < t\})}.$$

We can however easily check that

$$\lim_{m \to \infty} c(nm, t)^{1/m} = e^{2nt}$$

and the theorem follows. $\square$

Of course the same method gives the optimal version of the Herbort estimate (6.5) for arbitrary sublevel set:

$$\frac{|f(w)|^2}{K_\Omega(w, w)} \leq e^{-2nt} \int_{\{G_\Omega(\cdot, w) < t\}} |f|^2 d\lambda.$$

It is now the most interesting what happens with the right-hand side of (7.4) as $t \to -\infty$. For $n = 1$ we can write

$$G_\Omega(z, w) = \log|z - w| + \varphi(z),$$

where φ is harmonic in Ω. Denoting by M_t and m_t the supremum and infimum of φ over $\{G_\Omega(\cdot, w) < t\}$, respectively, we see that

$$\Delta(w, e^{t-M_t}) \subset \{G_\Omega(\cdot, w) < t\} \subset \Delta(w, e^{t-m_t})$$

and therefore

$$\lim_{t \to -\infty} e^{-2t} \lambda(\{G_\Omega(\cdot, w) < t\}) = \pi e^{-2\varphi(w)} = \frac{\pi}{c_\Omega(w)^2}.$$

We have thus obtained another proof of the Suita conjecture (7.2). Unlike the previous one which could have been presented entirely in dimension one, this one makes direct use of arbitrarily many complex variables to prove a one-dimensional result—the tensor power trick is crucial in this approach. Observe that this trick does not seem to work in another bound for the Bergman kernel (6.12)—there the constant

$$\left(\frac{1}{\sqrt{1 + a_{nm}^2}}\right)^{1/m}$$

increases to 1 as m increases to ∞, so in fact we get worse estimate than the original one.

In higher dimensions we have the following recent result from [31]:

Theorem 7.4 *Let Ω be bounded and hyperconvex. Then*

$$\lim_{t \to -\infty} e^{-2nt} \lambda(\{G_\Omega(\cdot, w) < t\}) = \lambda(I_\Omega^A(w)),$$

where

$$I_\Omega^A(w) = \left\{X \in \mathbb{C}^n : \overline{\lim_{\zeta \to 0}}(G_\Omega(w + \zeta X, w) - \log|\zeta|) < 0\right\}$$

is the Azukawa indicatrix of Ω at w.

Proof We may assume that $w = 0$. Write $G := G_{\Omega,0}$, $I_t := e^{-t}\{G < t\}$. By Zwonek [117] the function

$$A(X) = \overline{\lim_{\zeta \to 0}}(G(\zeta X) - \log|\zeta|)$$

is continuous on $\mathbb{C}^n$ and $\overline{\lim}$ is equal to $\lim$. Therefore

$$A(X) = \lim_{t \to -\infty}(G(e^t X) - t)$$

and by the Lebesgue bounded convergence theorem

$$\lim_{t \to -\infty} \lambda(I_t) = \lambda(\{A < 0\})$$

(if Ω is contained in $B(0, R)$ then so is I_t). $\qquad\square$

Combining this with Theorem 7.3 by approximation we thus obtain the following multidimensional version of the Suita conjecture:

Theorem 7.5 *For a pseudoconvex Ω and $w \in \Omega$ one has*

$$K_\Omega(w, w) \geq \frac{1}{\lambda(I_\Omega^A(w))}. \qquad \Box$$

It should be mentioned that recently Lempert [86] gave another proof of Theorem 7.3. He observed that considering the following pseudoconvex domain in $\mathbb{C}^{n+1}$

$$\{(z, \zeta) \in \Omega \colon G_\Omega(z, w) + \operatorname{Re} \zeta < 0\}$$

and using the result on log-plurisubharmoncity of sections of the Bergman kernel due to Maitani and Yamaguchi [89] for $n = 1$ and Berndtsson [6] for arbitrary n, one can get that the function $\log K_{\{G_\Omega(\cdot, w) < t\}}(w, w)$ is convex in t. For $r > 0$ with $B(w, r) \subset \Omega$ we have

$$\log K_{\{G_\Omega(\cdot, w) < t\}}(w, w) \leq -\log \lambda(B(w, re^t)),$$

and therefore the function

$$2nt + \log K_{\{G_\Omega(\cdot, w) < t\}}(w, w)$$

is convex and bounded from above on $(-\infty, 0]$, hence non-decreasing. We get

$$K_\Omega(w, w) \geq e^{2nt} K_{\{G_\Omega(\cdot, w) < t\}}(w, w) \geq \frac{e^{2nt}}{\lambda(\{G_\Omega(\cdot, w) < t\})},$$

since we can always take $f \equiv 1$ in (6.1). This gives another proof of the one-dimensional Suita conjecture, this time making crucial use of two complex variables.

Berndtsson and Lempert [12] very recently improved this method to obtain the Ohsawa–Takegoshi theorem with optimal constant as well. They use a stronger tool than log-plurisubharmoncity of sections of the Bergman kernel, namely Berndtsson's positivity of direct image bundles [8].

8 Suita conjecture for convex domains in $\mathbb{C}^n$

Theorems 7.3 and 7.5 seem to be especially interesting when Ω is convex. Then it is known, see [70], that the Lempert theory [85] implies that the Azukawa indicatrix $I_\Omega^A(w)$ is equal to the Kobayashi indicatrix

$$I_\Omega^K(w) = \{\varphi'(0) \colon \varphi \in \mathcal{O}(\Delta, \Omega), \ \varphi(0) = w\}.$$

We thus have the following estimate from [28]:

Theorem 8.1 *For $w \in \Omega \subset \mathbb{C}^n$, where Ω is a convex domain, we have*

$$K_\Omega(w, w) \geq \frac{1}{\lambda(I_\Omega^K(w))}.$$

$\square$

In this case, it turns out that a similar upper bound for the Bergman kernel also holds. We have the following result from [31]:

Theorem 8.2 *Under the assumptions of Theorem 8.1 we have*

$$K_\Omega(w, w) \leq \frac{4^n}{\lambda(I_\Omega^K(w))}.$$

If Ω is in addition symmetric with respect to w than the constant 4 above can be replaced with $16/\pi^2 = 1.621\ldots$

Proof Assume that $w = 0$ and let I be the interior of $I_\Omega^K(0)$. We will show that $I \subset 2\Omega$, then since I is balanced (that is $z \in I$ implies $\zeta z \in I$ for $\zeta \in \bar{\Delta}$) we will have

$$K_\Omega(0, 0) \leq K_{I/2}(0, 0) = \frac{1}{\lambda(I/2)} = \frac{4^n}{\lambda(I)}.$$

The proof that $I \subset 2\Omega$ will be similar to the proof of Proposition 1 in [95]. For $X = \varphi'(0) \in \bar{I}$ by L denote the complex line generated by X. Let a be the point from $L \cap \partial\Omega$ with the smallest distance to the origin, write it as $a = \zeta_0 X$. We want to show that $|X| \leq 2|a|$, that is that $|\zeta_0| \geq 1/2$.

Let H be the complex supporting hyperplane in $\mathbb{C}^n$ to Ω at a, that is $H \cap \Omega = \emptyset$ and $a \in H$. Without loss of generality we may assume that $H = \{z_n = a_n\}$. Let D be a half-plane in $\mathbb{C}$ containing the image of the projection of Ω to the nth variable and such that $a_n \in \partial D$. Then φ_n, the nth component of φ, belongs to $\mathcal{O}(\Delta, D)$ and $\varphi_n(0) = 0$. By the Schwarz lemma $|X_n| = |\varphi'_n(0)| \leq 2|a_n|$ which implies that $|\zeta_0| \geq 1/2$.

If Ω is in addition symmetric then as D we may take a strip instead of a half-plane and then $|\varphi'_n(0)| \leq (4/\pi)|a_n|$. $\square$

We have thus seen that for convex Ω the biholomorphically invariant function

$$F_\Omega(w) := (K_\Omega(w, w)\lambda(I_\Omega^K(w)))^{1/n}$$

satisfies

$$1 \leq F_\Omega \leq 4.$$

The lower bound was obtained using the $\bar{\partial}$-equation whereas the proof of the upper bound was relatively elementary. The lower bound is optimal—for example if Ω is balanced with respect to w then we have equality—and it would be interesting to find an optimal upper bound. It is in fact not so trivial to prove that we may at all have

$F_\Omega(w) > 1$. This was done in [31] and [32] where F_Ω was computed for certain complex convex ellipsoids and some w. Here are two results:

Theorem 8.3 *For* $\Omega = \{z \in \mathbb{C}^n : |z_1| + \cdots + |z_n| < 1\}$ *and* $w = (b, 0, \ldots, 0)$, *where* $0 \le b < 1$, *one has*

$$K_\Omega(w)\lambda(I_\Omega^K(w)) = 1 + (1-b)^{2n}\frac{(1+b)^{2n} - (1-b)^{2n} - 4nb}{4nb(1+b)^{2n}}$$

$$= 1 + \frac{(1-b)^{2n}}{(1+b)^{2n}} \sum_{j=1}^{n-1} \frac{1}{2j+1}\binom{2n-1}{2j}b^{2j}.$$

The proof uses the formula for the Bergman kernel for this ellipsoid

$$K_\Omega((b, 0, \ldots, 0)) = \frac{2n-1}{4\pi\omega b}((1-b)^{-2n} - (1+b)^{-2n}),$$

where $\omega = \lambda(\{z \in \mathbb{C}^{n-1} : |z_1| + \cdots + |z_{n-1}| < 1\})$, obtained from the deflation method of Boas–Fu–Straube [33]. The main part of the proof was to compute $\lambda(I_\Omega^K(w))$. For that the formula of Jarnicki–Pflug–Zeinstra [71] for geodesics in convex complex ellipsoids (which is based on Lempert's theory [85]) was used. Here are the resulting graphs of $F_\Omega(b, 0, \ldots, 0)$ for $n = 2, 3, \ldots, 6$:

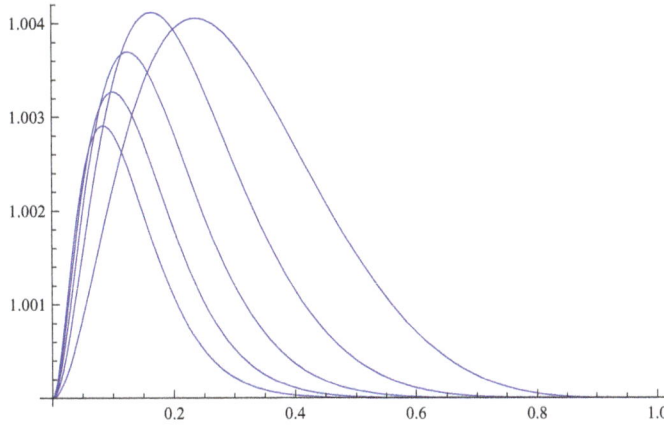

Theorem 8.4 *For* $m \ge 1/2$ *set* $\Omega_m := \{|z_1|^{2m} + |z_2|^2 < 1\}$ *and* $w = (b, 0)$ *where* $0 \le b < 1$. *Then*

$$\lambda(I_{\Omega_m}((b, 0))) = \pi^2\Bigg[-\frac{m-1}{2m(3m-2)(3m-1)}b^{6m+2} - \frac{3(m-1)}{2m(m-2)(m+1)}b^{2m+2}$$

$$+ \frac{m}{2(m-2)(3m-2)}b^6 + \frac{3m}{3m-1}b^4 - \frac{4m-1}{2m}b^2 + \frac{m}{m+1}\Bigg].$$

Some computations leading to this formula were done with the help of *Mathematica*. The Kobayashi function for this ellipsoid was computed implicitly by Blank–Fan–Klein–Krantz–Ma–Pang [13] (explicitly up to solving a real equation which is a polynomial one of degree $2m$ if m is an integer) and this had only sufficed for numerical computations of $\lambda(I_{\Omega_m}^K(w))$. It turns out however that just the indicatrix $I_{\Omega_m}^K(w)$ and its volume can be described with explicit although rather complicated formulas. Here is the graph of $F_{\Omega_m}(b, 0)$ for $m = 4, 8, 16, 32, 64$ and 128:

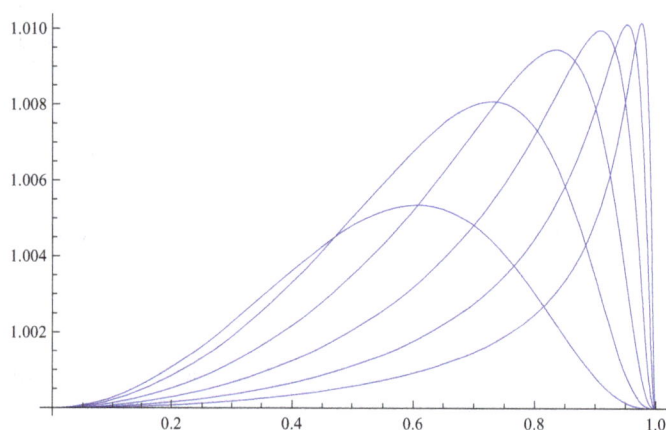

In this particular case all values of $F_{\Omega_m}(w)$ are attained for $w = (b, 0)$, $0 < b < 1$. One can compute numerically that

$$\sup_{m \geq 1/2} \sup_{\Omega_m} F_{\Omega_m} = 1.010182\ldots$$

and this is the highest value of F_Ω for convex Ω in any dimension we have been able to obtain so far. It seems that the lower bound given by Theorem 8.1 is very accurate.

9 Mahler conjecture and Bourgain–Milman inequality

Let K be a convex symmetric (that is $K = -K$) body (that is K is compact and has non-empty interior) in $\mathbb{R}^n$. Its *dual* is defined by

$$K' := \{y \in \mathbb{R}^n : x \cdot y \leq 1 \text{ for all } x \in K\},$$

where $\cdot$ denotes the inner product in $\mathbb{R}^n$. The number

$$\lambda(K)\lambda(K')$$

is called the *Mahler volume* of K. One can show that it is independent of linear transformations of $\mathbb{R}^n$ and of the choice of the inner product. It is thus an invariant of the n-dimensional real Banach space whose unit ball is K. Santaló [105] showed that the Mahler volume is maximized for balls (it was earlier proved by Blaschke in dimensions 2 and 3) and by Saint–Raymond [104] these are the only maximizers (up to linear transformations). For a proof of the Blaschke–Santaló inequality using the $\bar{\partial}$-equation see [41].

Mahler [88] conjectured that the Mahler volume is maximized by cubes. He proved it in dimension 2 and the problem still remains open in higher dimensions. This, together with the Blaschke–Santaló inequality, would mean that the Mahler volume is biggest for the roundest convex symmetric bodies and smallest for the least round. One of the difficulties with the Mahler conjecture is that, if true, cubes cannot be the only minimizers, even up to linear transformations. The other candidates are the so called *Hansen–Lima bodies* [60]: in $\mathbb{R}$ these are symmetric closed intervals and in higher dimensions they are produced by taking either products of lower dimensional Hansen–Lima bodies or a dual. This way we do not get anything new in $\mathbb{R}^2$, since the dual of $[-1, 1]^2$ is the linearly equivalent rhombus $\{|x_1| + |x_2| \leq 1\}$. However, already in dimension 3 the dual of the unit cube $[-1, 1]^3$ is the octahedron $\{|x_1| + |x_2| + |x_3| < 1\}$ and they are not linearly equivalent. These two are the only Hansen–Lima bodies in $\mathbb{R}^3$ and there are more in higher dimensions. It is conjectured that Hansen–Lima bodies are the only minimizers of the Mahler volume (up to linear transformations).

An important lower bound for the Mahler volume is the Bourgain–Milman inequality [34]:

Theorem 9.1 *There exists an absolute constant $c > 0$ such that for a symmetric convex body K in $\mathbb{R}^n$ one has*

$$\lambda(K)\lambda(K') \geq c^n \frac{4^n}{n!}. \tag{9.1}$$

Since the Mahler volume of a cube in $\mathbb{R}^n$ is equal to $4^n/n!$, the Mahler conjecture is equivalent to (9.1) with $c = 1$. The original proof from [34] was qualitative, it did not give any particular value of c. So far the best constant in (9.1) was obtained by Kuperberg [84] who proved it with $c = \pi/4$. Recently Nazarov [93] gave a different proof of the Bourgain–Milman inequality and although he obtained a worse constant than Kuperberg, namely $c = (\pi/4)^3$, his proof was very interesting from our point of view because he used several complex variables and Hörmander's estimate. In [28] it was shown that Theorem 8.1 can be used in Nazarov's approach instead but of course Hörmander's estimate is hidden there.

Before we present Nazarov's proof of Theorem 9.1, let us look at his equivalent formulation of the Mahler conjecture as a problem in several complex variables. For $u \in L^2(K')$ and its Fourier transform $\widehat{u} \in \mathcal{O}(\mathbb{C}^n)$ by the Schwarz inequality and the Plancherel formula we have

$$|\widehat{u}(0)|^2 = \left| \int_{K'} u \, d\lambda \right|^2 \leq \lambda(K')||u||^2_{L^2(K')} = (2\pi)^{-n}\lambda(K')||\widehat{u}||^2_{L^2(\mathbb{R}^n)}$$

and the equality holds if u is the characteristic function of K'. Therefore

$$\lambda(K') = (2\pi)^n \sup_{f \in \mathcal{P}} \frac{|f(0)|^2}{||f||^2_{L^2(\mathbb{R}^n)}}, \tag{9.2}$$

where

$$\mathcal{P} = \{\widehat{u} : u \in L^2(K')\}$$

is a family of entire holomorphic functions. In fact, using the Paley–Wiener theorem one can completely characterize the class $\mathcal{P}$ (see e.g. [103] for details): it consists of those $f \in \mathcal{O}(\mathbb{C}^n)$ that are of exponential growth (that is $|f(z)| \leq Ce^{C|z|}$ for some constant C) and such that

$$|f(iy)| \leq Ce^{q_K(y)}, \quad y \in \mathbb{R}^n,$$

where q_K is the Minkowski function for K (that is the norm in $\mathbb{R}^n$ with unit ball K). The usefulness of the formula for the volume of the dual (9.2) is that K' itself does not appear on the right-hand side. Therefore the Mahler conjecture is equivalent to finding $f \in \mathcal{P}$ such that $f(0) = 1$ and

$$\int_{\mathbb{R}^n} |f(x)|^2 d\lambda(x) \leq \frac{n!\pi^n}{2^n}\lambda(K).$$

Nazarov, instead of constructing a holomorphic function on the entire $\mathbb{C}^n$, considered the convex tube in $\mathbb{C}^n$ defined by K:

$$T_K := \text{int } K + i\mathbb{R}^n.$$

He proved the following bounds for the Bergman kernel in T_K:

$$K_{T_K}(0,0) \leq \frac{n!}{\pi^n}\frac{\lambda(K')}{\lambda(K)} \tag{9.3}$$

and

$$K_{T_K}(0,0) \geq \left(\frac{\pi}{4}\right)^{2n}\frac{1}{(\lambda(K))^2}. \tag{9.4}$$

Combining them we get (9.1) with $c = (\pi/4)^3$. Note that (9.4) follows immediately from Theorem 8.1 and the following:

Proposition 9.2 *For a convex symmetric body K in $\mathbb{R}^n$ we have*

$$I_{T_K}^K(0) \subset \frac{4}{\pi}(K + iK).$$

Proof Let Φ be a conformal mapping from the strip $(-1, 1) + i\mathbb{R}$ to Δ such that $\Phi(0) = 0$, then $|\Phi'(0)| = \pi/4$. Fix $y \in K'$, then

$$F(z) := \Phi(z \cdot y) \in \mathcal{O}(T_K, \Delta)$$

satisfies $F(0) = 0$. For $X = \varphi'(0) \in I_{T_K}^K(0)$ by the Schwarz lemma we have $|(F \circ \varphi)'(0)| \leq 1$ and therefore $|X \cdot y| \leq 4/\pi$. This means that

$$I_{T_K}^K(0) \subset \frac{4}{\pi}\{z \in \mathbb{C}^n : |z \cdot y| \leq 1 \text{ for all } y \in K'\} \subset \frac{4}{\pi}(K'' + iK'')$$

and the proposition follows since $K'' = K$. $\square$

For smooth strongly convex K Lempert's theory [85] can be used to obtain more precise description of $I_{T_K}^K(0)$ in terms of the Gauss mapping of ∂K, see [28].

As shown in [93] the upper bound (9.3) follows easily from the formula for the Bergman kernel in convex tube domains due to Rothaus [102] (see also [68]):

$$K_{T_K}(z, w) = \frac{1}{(2\pi)^n} \int_{\mathbb{R}^n} \frac{e^{i(z-\bar{w})\cdot y}}{J_K(y)} d\lambda(y),$$

where

$$J_K(y) = \int_K e^{-2x\cdot y} d\lambda(x).$$

Fix $y \in \mathbb{R}^n$ and $\tilde{x} \in K$. Then, since K is symmetric,

$$J_K(y) \geq 2^{-n} \int_K e^{-(x+\tilde{x})\cdot y} d\lambda(x) \geq 2^{-n}\lambda(K)e^{-\tilde{x}\cdot y}.$$

Minimizing the right-hand side over $\tilde{x}$ we get

$$J_K(y) \geq 2^{-n} e^{q_{K'}(y)}.$$

Since for any convex body K one has

$$\int_{\mathbb{R}^n} e^{-q_K} d\lambda = \int_{\mathbb{R}^n} \int_{q_K(y)}^\infty e^{-t} dt \, d\lambda(y) = \int_0^\infty e^{-t}\lambda(\{q_K < t\}) dt = n!\lambda(K),$$

the upper bound (9.3) follows.

10 Isoperimetric inequalities and symmetrization

One of the interesting open problems is whether in the lower bound for the Bergman kernel (7.4) the right-hand side is monotone in t. This would mean in particular that the best bound is obtained when $t \to -\infty$, that is that Theorem 7.5 is the optimal version of this estimate. We start with the following result from [31] showing that this is indeed the case for $n = 1$:

Theorem 10.1 *Assume that $w \in \Omega \subset \mathbb{C}$. Then the function*

$$(-\infty, 0] \ni t \longmapsto e^{-2t}\lambda(\{G_\Omega(\cdot, w) < t\})$$

is non-decreasing.

Proof With the notation $G := G_\Omega(\cdot, w)$ set

$$f(t) := \log \lambda(\{G < t\}) - 2t.$$

It is enough to show that if t is a regular value of G then $f'(t) \geq 0$. By the co-area formula

$$\lambda(\{G < t\}) = \int_{-\infty}^{t} \int_{\{G=s\}} \frac{d\sigma}{|\nabla G|} \, ds$$

and therefore by the Schwarz inequality

$$\frac{d}{dt} \lambda(\{G < t\}) = \int_{\{G=t\}} \frac{d\sigma}{|\nabla G|} \geq \frac{\sigma(\{G=t\})^2}{\int_{\{G=t\}} |\nabla G| d\sigma}.$$

We have

$$\int_{\{G=t\}} |\nabla G| d\sigma = \int_{\{G=t\}} \frac{\partial G}{\partial n} d\sigma = \int_{\{G<t\}} \Delta G = 2\pi$$

and by the isoperimetric inequality

$$\sigma(\{G=t\})^2 \geq 4\pi \lambda(\{G < t\}).$$

It follows that

$$f'(t) = \frac{\int_{\{G=t\}} \dfrac{d\sigma}{|\nabla G|}}{\lambda(\{G < t\})} - 2 \geq 0.$$

$\square$

Note that the proof also shows that the problem whether for pseudoconvex $\Omega \subset \mathbb{C}^n$ the function

$$(-\infty, 0] \ni t \longmapsto e^{-2nt} \lambda(\{G_\Omega(\cdot, w) < t\}) \tag{10.1}$$

is non-decreasing is equivalent to the following "pluripolar isoperimetric inequality": if Ω is bounded, smooth and strongly pseudoconvex in $\mathbb{C}^n$ then for $w \in \Omega$ one has

$$\int_{\partial\Omega} \frac{d\sigma}{|\nabla G_\Omega(\cdot, w)|} \geq 2\lambda(\Omega).$$

Similarly as in Lempert's proof of Theorem 7.3, the monotonicity of (10.1) would follow if we knew that the function

$$(-\infty, 0] \ni t \longmapsto \log \lambda(\{G_\Omega(\cdot, w) < t\})$$

was convex. This was conjectured in [31] but Fornæss [55] found a counterexample to that already for $n = 1$.

The method of proof of Theorem 10.1 was in fact inspired by the proof of a symmetrization result for the Laplacian due to Talenti [111]. For a measurable subset A in $\mathbb{R}^n$ its *Schwarz symmetrization* (or *rearrangement*) A^* is the ball centered at the origin such that $\lambda(A^*) = \lambda(A)$. For a nonnegative measurable function f defined on a measurable subset A of $\mathbb{R}^n$ its Schwarz symmetrization f^* is the radially symmetric (that is $f^*(x)$ depends only on $|x|$) function defined on A^* which is non-increasing in radius and such that $\lambda(\{f^* > t\}) = \lambda(\{f > t\})$ for every real t. If f is nonpositive than we set $f^* := -(-f)^*$ or equivalently require that f^* is non-decreasing in radius and the volumes of sublevel (instead of superlevel) sets are the same. One of the useful properties of rearrangements is that they preserve the L^p-norms, or more generally

$$\int_{\Omega^*} \gamma(|f^*|)\, d\lambda = \int_{\Omega} \gamma(|f|)\, d\lambda$$

for any increasing γ and f either nonpositive or nonnegative. For an introduction to rearrangements we refer to [36].

Talenti [111] proved the following:

Theorem 10.2 *Let Ω be a bounded regular domain in $\mathbb{R}^n$ and let u be a (possibly weak) solution to the following Dirichlet problem*

$$\begin{cases} \Delta u = f \geq 0 & \text{in } \Omega \\ u = 0 & \text{on } \partial\Omega \end{cases}.$$

If v solves

$$\begin{cases} \Delta v = f^* & \text{in } \Omega^* \\ v = 0 & \text{on } \partial\Omega^* \end{cases},$$

then $v \leq u^*$ in Ω^*.

Proof By approximation we may assume that u is smooth and strongly subharmonic. By the Hardy–Littlewood inequality for $t \leq 0$ we have

$$\int_{\{u < t\}} f\, d\lambda \leq \int_{\{u^* < t\}} f^*\, d\lambda = \int_{B(0,r)} \Delta v\, d\lambda = n\omega_n r^{n-1} \gamma'(r),$$

where r is such that $\{u^* < t\} = B(0,r)$, $v(x) = \gamma(|x|)$, ω_n is the volume of the unit ball in $\mathbb{R}^n$, and the last equality follows from the fact that

$$\Delta v = \gamma'' + (n-1)\frac{\gamma'}{r} = r^{1-n}\frac{d}{dr}(r^{n-1}\gamma').$$

On the other hand, if t is a regular value of u then by the Schwarz inequality

$$\int_{\{u<t\}} f\, d\lambda = \int_{\{u=t\}} |\nabla u|\, d\sigma \geq \frac{\sigma(\{u=t\})^2}{\displaystyle\int_{\{u=t\}} \frac{d\sigma}{|\nabla u|}}.$$

By the isoperimetric inequality

$$\sigma(\{u=t\}) \geq n\omega_n^{1/n}\lambda(\{u<t\})^{1-1/n}$$

and by the co-area formula

$$\int_{\{u=t\}} \frac{d\sigma}{|\nabla u|} = \frac{d}{dt}\lambda(\{u<t\}).$$

Therefore

$$\int_{\{u<t\}} f\, d\lambda \geq n^2\omega_n^{2/n}\frac{\lambda(\{u<t\})^{2-2/n}}{\dfrac{d}{dt}\lambda(\{u<t\})}.$$

Write $u^*(x) = \eta(|x|)$. Since $\{u^* < t\} = B(0,r)$, we have $t = \eta(r)$ and $\lambda(\{u<t\}) = \omega_n r^n$. Therefore

$$\int_{\{u<t\}} f\, d\lambda \geq n\omega_n r^{n-1}\eta'(r)$$

and it follows that $\eta' \leq \gamma'$. Since $\eta(R) = \gamma(R) = 0$, where $\Omega^* = B(0,R)$, we obtain that $\eta \geq \gamma$.   □

For a corresponding symmetrization result for the real Monge–Ampère equation one has to symmetrize convex u with respect to a different measure. For a bounded convex domain Ω in $\mathbb{R}^n$ its quermassintegrals $V_m(\Omega)$, $m = 0, 1, \ldots, n$, are defined by the formula

$$\lambda(\Omega + t\mathbb{B}) = \sum_{m=0}^{n} \binom{n}{m} V_{n-m}(\Omega)t^m,$$

where $\mathbb{B}$ is the unit ball in $\mathbb{R}^n$ and $t \geq 0$. Then $V_n(\Omega) = \lambda(\Omega)$, $V_{n-1}(\Omega) = \sigma(\partial\Omega)/n$ (if Ω is smooth) and $V_0(\Omega) = \omega_n$. We also have $V_m(B(0,r)) = \omega_n r^m$. Alexandrov–Fenchel inequalities state that the expression

$$(V_m(\Omega)/\omega_n)^{1/m}$$

is non-increasing in m and we have equality at any stage only for balls.

If Ω is in addition smooth then $V_m(\Omega)$ can be expressed in terms of an integral over $\partial\Omega$ of a proper curvature of $\partial\Omega$. If $\kappa_1, \ldots, \kappa_{n-1}$ are the principal curvatures of $\partial\Omega$ then the mth mean curvature of $\partial\Omega$, $m = 1, \ldots, n-1$, is defined by

$$H_m := \sum_{1 \leq i_1 < \cdots < i_m \leq n-1} \kappa_{i_1} \ldots \kappa_{i_m}.$$

For $m = 0$ we set $H_0 \equiv 1$. Then H_1 is the mean curvature and H_{n-1} the Gauss curvature of $\partial\Omega$. Then for $m = 0, 1, \ldots, n-1$ we have

$$V_m(\Omega) = \frac{1}{n\binom{n-1}{m}} \int_{\partial\Omega} H_{n-m-1} d\sigma.$$

We refer to [35] and [106] for more details.

By $\tilde{}$ we will denote the symmetrization with respect to V_1 instead of the Lebesgue measure λ. Note that by the Alexandrov–Fenchel inequalities we have $\Omega^* \subset \widetilde{\Omega}$ and $\widetilde{u} \leq u^*$ for negative convex u. We have the following result for the real Monge–Ampère equation due to Talenti [112] in dimension 2 and Tso [113] in the general case.

Theorem 10.3 *Let Ω be a bounded convex domain. Assume that u is a (possibly weak) convex solution to the Dirichlet problem*

$$\begin{cases} \det D^2 u = f \geq 0 & in \ \ \Omega \\ u = 0 & on \ \ \partial\Omega \end{cases}.$$

Extend f^* by 0 from Ω^* to $\widetilde{\Omega}$. If convex v solves

$$\begin{cases} \det D^2 v = f^* & in \ \ \widetilde{\Omega} \\ v = 0 & on \ \ \partial\widetilde{\Omega} \end{cases},$$

then $v \leq \widetilde{u}$ in $\widetilde{\Omega}$.

Proof By approximation we may assume that u is smooth and strongly convex. Similarly as in the proof of Theorem 10.2 we have

$$\int_{\{u < t\}} f \, d\lambda \leq \int_{\{u^* < t\}} f^* d\lambda \leq \int_{\{\widetilde{u} < t\}} f^* d\lambda = \int_{B(0,r)} \det D^2 v = \omega_n (\gamma'(r))^n,$$

where r is such that $\{\widetilde{u} < t\} = B(0, r)$ and $v(x) = \gamma(|x|)$, so that

$$\det D^2 v = r^{1-n} (\gamma')^{n-1} \gamma'' = \frac{1}{n} r^{1-n} \frac{d}{dr}((\gamma')^n).$$

On the other hand for the regular value t of u we have by the Hölder inequality

$$\int_{\{u<t\}} f\,d\lambda = \int_{\{u=t\}} |\nabla u|^n H_{n-1}\,d\sigma \geq \frac{\left(\int_{\{u=t\}} H_{n-1}\,d\sigma\right)^{n+1}}{\left(\int_{\{u=t\}} \frac{H_{n-1}}{|\nabla u|}\,d\sigma\right)^n}.$$

We have

$$\int_{\{u=t\}} H_{n-1}\,d\sigma = n\omega_n$$

and by Reilly [101]

$$\int_{\{u=t\}} \frac{H_{n-1}}{|\nabla u|}\,d\sigma = \frac{1}{n-1}\frac{d}{dt}\int_{\{u=t\}} H_{n-2}\,d\sigma = n\frac{d}{dt}V_1(\{u<t\}). \qquad (10.2)$$

If $\tilde{u}(x) = \eta(|x|)$ then, since $\{\tilde{u} < t\} = B(0,r)$, we have $t = \eta(r)$. We will obtain

$$\int_{\{u<t\}} f\,d\lambda \geq \omega_n(\eta'(r))^n,$$

and thus $\eta' \leq \gamma'$. Since $\eta(R) = \gamma(R) = 0$, where $\tilde{\Omega} = B(0,R)$, we get $\eta \geq \gamma$.   □

It would be very desirable to prove a similar result for the complex Monge–Ampère equation. This would in particular immediately imply the following important estimate of Kołodziej [81] (see also [82]):

Theorem 10.4 *Let Ω be a bounded hyperconvex domain in $\mathbb{C}^n$ and let be a solution to the following Dirichlet problem*

$$\begin{cases} u \in PSH(\Omega) \cap C(\Omega) \\ (dd^c u)^n = f\,d\lambda & in\ \Omega \\ u = 0 & on\ \partial\Omega \end{cases}.$$

Then for every $p > 1$ one has

$$\sup_{\Omega} |u| \leq C\|f\|_{L^p(\Omega)}^{1/n},$$

where C is a constant depending on n, p and the diameter of Ω.

Similarly as for convex domains, for a smooth pseudoconvex Ω one can consider the Levi principal curvatures of the boundary $\lambda_1, \ldots, \lambda_{n-1}$ and define the mth complex mean curvature K_m similarly as H_m, so that $K = K_{n-1}$ is the Levi curvature of the boundary. See [92] and [90] for basic results on complex mean curvatures. If one tries to repeat the method of the proof of Theorem 10.3 then two problems appear: first is the lack of complex counterparts of the Alexandrov–Fenchel inequalities and secondly it is not clear what the Reilly formula (10.2) should look like in the complex case. It is also not at all clear what the right symmetrization $\sim$ should be now. One of interesting conjectures that arise (although not sufficient to prove a symmetrization result for the complex Monge–Ampère equation), is the following: for a bounded smooth strongly pseudoconvex Ω in $\mathbb{C}^n$ the following complex isoperimetric inequality holds:

$$\int_{\partial\Omega} K\, d\sigma \geq 2n\sqrt{\omega_{2n}\lambda(\Omega)}$$

with equality exactly for balls.

References

1. Bedford, E., Taylor, B.A.: The Dirichlet problem for a complex Monge–Ampère equation. Invent. Math. **37**, 1–44 (1976)
2. Bedford, E., Taylor, B.A.: A new capacity for plurisubharmonic functions. Acta Math. **149**, 1–41 (1982)
3. Bedford, E., Demailly, J.-P.: Two counterexamples concerning the pluri-complex Green function in $\mathbb{C}^n$. Indiana Univ. Math. J. **37**, 865–867 (1988)
4. Berndtsson, B.: The extension theorem of Ohsawa–Takegoshi and the theorem of Donnelly–Fefferman. Ann. Inst. Fourier **46**, 1083–1094 (1996)
5. Berndtsson, B.: Weighted estimates for the $\bar{\partial}$-equation. In: Complex Analysis and Geometry, Columbus, 1999. Ohio State University Mathematical Research Institute, vol. 9, pp. 43–57. Walter de Gruyter, Berlin (2001)
6. Berndtsson, B.: Subharmonicity properties of the Bergman kernel and some other functions associated to pseudoconvex domains. Ann. Inst. Fourier **56**, 1633–1662 (2006)
7. Berndtsson, B.: L^2-estimates for the d-equation and Witten's proof of the Morse inequalities. Ann. Fac. Sci. Toulouse Math. **16**, 773–797 (2007)
8. Berndtsson, B.: Curvature of vector bundles associated to holomorphic fibrations. Ann. Math. **169**, 531–560 (2009)
9. Berndtsson, B.: An introduction to things $\bar{\partial}$. In: Analytic and algebraic geometry, IAS/Park City Mathematics Series, vol 17, pp. 7–76. American Mathematical Society (2010)
10. Berndtsson, B.: The openness conjecture for plurisubharmonic functions.
11. Berndtsson, B.: Private communication (2014)
12. Berndtsson, B., Lempert, L.: A proof of the Ohsawa–Takegoshi theorem with sharp estimates.

13. Blank, B.E., Fan, D.S., Klein, D., Krantz, S.G., Ma, D., Pang, M.-Y.: The Kobayashi metric of a complex ellipsoid in $\mathbb{C}^2$. Exp. Math. **1**, 47–55 (1992)
14. Błocki, Z.: Estimates for the complex Monge–Ampère operator. Bull. Pol. Acad. Sci. Math. **41**, 151–157 (1993)
15. Błocki, Z.: The complex Monge–Ampère operator in hyperconvex domains. Ann. Scuola Norm. Sup. Pisa **23**, 721–747 (1996)
16. Błocki, Z.: Interior regularity of the complex Monge–Ampère equation in convex domains. Duke Math. J. **105**, 167–181 (2000)
17. Błocki, Z.: Equilibrium measure of a product subset of $\mathbb{C}^n$. Proc. Am. Math. Soc. **128**, 3595–3599 (2000)

18. Błocki, Z.: The $C^{1,1}$ regularity of the pluricomplex Green function. Mich. Math. J. **47**, 211–215 (2000)
19. Błocki, Z.: Regularity of the pluricomplex Green function with several poles. Indiana Univ. Math. J. **50**, 335–351 (2001)
20. Błocki, Z.: The complex Monge–Ampère operator in pluripotential theory. In: Lecture Notes (2002).

21. Błocki, Z.: A note on the Hörmander, Donnelly–Fefferman, and Berndtsson L^2-estimates for the $\bar{\partial}$-operator. Ann. Pol. Math. **84**, 87–91 (2004)
22. Błocki, Z.: On the definition of the Monge–Ampère operator in $\mathbb{C}^2$. Math. Ann. **328**, 415–423 (2004)
23. Błocki, Z.: The Bergman metric and the pluricomplex Green function. Trans. Am. Math. Soc. **357**, 2613–2625 (2005)
24. Błocki, Z.: The domain of definition of the complex Monge–Ampère operator. Am. J. Math. **128**, 519–530 (2006)
25. Błocki, Z.: Some estimates for the Bergman kernel and metric in terms of logarithmic capacity. Nagoya Math. J. **185**, 143–150 (2007)
26. Błocki, Z.: On the Ohsawa–Takegoshi extension theorem. Univ. Lag. Acta Math. **50**, 53–61 (2012)
27. Błocki, Z.: Suita conjecture and the Ohsawa–Takegoshi extension theorem. Invent. Math. **193**, 149–158 (2013)
28. Błocki, Z.: A lower bound for the Bergman kernel and the Bourgain–Milman inequality. In: GAFA Seminar Notes. Lecture Notes in Mathematics. Springer, New York (2014, to appear)
29. Błocki, Z.: Estimates for $\bar{\partial}$ and optimal constants. In: Proceedings of the Abel Symposium 2013. Springer, New York (2014, to appear)
30. Błocki, Z., Pflug, P.: Hyperconvexity and Bergman completeness. Nagoya Math. J. **151**, 221–225 (1998)
31. Błocki, Z., Zwonek, W.: Estimates for the Bergman kernel and the multidimensional Suita conjecture.

32. Błocki, Z., Zwonek, W.: On the Suita conjecture for some convex ellipsoids in $\mathbb{C}^2$.
33. Boas, H.P., Fu, S., Straube, E.J.: The Bergman kernel function: explicit formulas and zeroes. Proc. Am. Math. Soc. **127**, 805–811 (1999)
34. Bourgain, J., Milman, V.: New volume ratio properties for convex symmetric bodies in $\mathbb{R}^n$. Invent. Math. **88**, 319–340 (1987)
35. Burago, Y.D., Zalgaller, V.A.: Geometric Inequalities. Springer-Verlag, New York (1988)
36. Burchard, A.: A short course on rearrangement inequalities. In: Lecture Notes (2009).

37. Caffarelli, L., Kohn, J.J., Nirenberg, L., Spruck, J.: The Dirichlet problem for non-linear second order elliptic equations II: complex Monge–Ampère, and uniformly elliptic equations. Commun. Pure Appl. Math. **38**, 209–252 (1985)
38. Chen, B.Y.: Completeness of the Bergman metric on non-smooth pseudoconvex domains. Ann. Pol. Math. **71**, 241–251 (1999)
39. Chen, B.Y.: A remark on the Bergman completeness. Complex Var. Theory Appl. **42**, 11–15 (2000)
40. Chen, B.Y.: A simple proof of the Ohsawa–Takegoshi extension theorem.
41. Cordero-Erausquin, D.: Santaló's inequality on $\mathbb{C}^n$ by complex interpolation. C. R. Math. Acad. Sci. Paris **334**, 767–772 (2002)
42. Demailly, J.-P.: Estimations L^2 pour l'opérateur $\bar{\partial}$ d'un fibré vectoriel holomorphe semi-positif au-dessus d'une variété kählérienne complète. Ann. Sci. École Norm. Sup. **15**, 457–511 (1982)
43. Demailly, J.-P.: Mesures de Monge–Ampère et mesures plurisousharmoniques. Math. Z. **194**, 519–564 (1987)
44. Demailly, J.-P.: Nombres de Lelong généralisés, théorèmes d'intégralité et d'analyticité. Acta Math. **159**, 153–169 (1987)
45. Demailly, J.-P.: Regularization of closed positive currents and intersection theory. J. Algebraic Geom. **1**, 361–409 (1992)
46. Demailly, J.P., Kollár, J.: Semicontinuity of complex singularity exponents and Kähler–Einstein metrics on Fano orbifolds. Ann. Sci. École Norm. Sup. **34**, 525–556 (2001)
47. Demailly, J.-P., Peternell, T., Schneider, M.: Pseudo-effective line bundles on compact Khler manifolds. Int. J. Math. **12**, 689–741 (2001)
48. Diederich, K., Fornæss, J.E.: Pseudoconvex domains: bounded plurisubharmonic exhaustion functions. Invent. Math. **39**, 129–141 (1977)

49. Diederich, K., Ohsawa, T.: An estimate for the Bergman distance on pseudoconvex domains. Ann. Math. **141**, 181–190 (1995)
50. Dinew, Ż.: On the Bergman representative coordinates. Sci. China Math. **54**, 1357–1374 (2011)
51. Dinew, Ż.: An example for the holomorphic sectional curvature of the Bergman metric. Ann. Pol. Math. **98**, 147–167 (2010)
52. Donnelly, H., Fefferman, C.: L^2-cohomology and index theorem for the Bergman metric. Ann. Math. **118**, 593–618 (1983)
53. Edigarian, A.: On the product property of the pluricomplex Green function. Proc. Am. Math. Soc. **125**, 2855–2858 (1997)
54. Favre, C., Jonsson, M.: Valuations and multiplier ideals. J. Am. Math. Soc. **18**, 655–684 (2005)
55. Fornæss, J.E.: Private communication (2014)
56. Guan, B.: The Dirichlet problem for complex Monge–Ampère equations and regularity of the pluricomplex Green function. Commun. Anal. Geom. **6**, 687–703 (1998) (correction: ibid. 8, 2000, 213–218)
57. Guan, Q.A., Zhou, X.Y.: A solution of an L^2 extension problem with optimal estimate and applications. Ann. Math. (2014, to appear).
58. Guan, Q.A., Zhou, X.Y.: Strong openness conjecture for plurisubharmonic functions.
59. Guan, Q.A., Zhou, X.Y., Zhu, L.F.: On the Ohsawa–Takegoshi L^2 extension theorem and the Bochner–Kodaira identity with non-smooth twist factor. J. Math. Pures Appl. **97**, 579–601 (2012)
60. Hansen, A.B., Lima, Å.: The structure of finite-dimensional Banach spaces with the 3.2. intersection property. Acta Math. **146**, 1–23 (1981)
61. Harrington, P.S.: The order of plurisubharmonicity on pseudoconvex domains with Lipschitz boundaries. Math. Res. Lett. **14**, 485–490 (2007)
62. Herbort, G.: The Bergman metric on hyperconvex domains. Math. Z. **232**, 183–196 (1999)
63. Herbort, G.: The pluricomplex Green function on pseudoconvex domains with a smooth boundary. Int. J. Math. **11**, 509–522 (2000)
64. Hiep, P.H.: The weighted log canonical threshold.
65. Hörmander, L.: L^2 estimates and existence theorems for the $\bar{\partial}$ operator. Acta Math. **113**, 89–152 (1965)
66. Hörmander, L.: An Introduction to Complex Analysis in Several Variables. North Holland, Amsterdam (1991)
67. Hörmander, L.: Notions of Convexity. Birkhäuser, Basel (1994)
68. Hsin, C.-I.: The Bergman kernel on tube domains. Rev. Un. Mat. Argent. **46**, 23–29 (2005)
69. Jarnicki, M., Pflug, P.: Invariant pseudodistances and pseudometrics—completeness and product property. Ann. Pol. Math. **55**, 169–189 (1991)
70. Jarnicki, M., Pflug, P.: Invariant Distances and Metrics in Complex Analysis. Walter de Gruyter, Berlin (1993)
71. Jarnicki, M., Pflug, P., Zeinstra, R.: Geodesics for convex complex ellipsoids. Ann. Scuola Norm. Sup. Pisa **20**, 535–543 (1993)
72. Jucha, P.: Bergman completeness of Zalcman type domains. Stud. Math. **163**, 71–82 (2004)
73. Kerzman, N., Rosay, J.-P.: Fonctions plurisousharmoniques dexhaustion bornées et domaines taut. Math. Ann. **257**, 171–184 (1981)
74. Kim, D.: A remark on the approximation of plurisubharmonic functions. C. R. Math. Acad. Sci. Paris **352**, 387–389 (2014)
75. Kiselman, C.O.: Densité des fonctions plurisousharmoniques. Bull. Soc. Math. France **107**, 295–304 (1979)
76. Kiselman, C.O.: Sur la définition de lopérateur de MongeAmpère complexe. Analyse Complexe. In: Proceedings of the Journées Fermat Journées SMF, Toulouse 1983. Lecture Notes in Mathematics, vol. 1094, pp. 139–150. Springer, New York (1984)
77. Kiselman, C.O.: La teoremo de Siu por abstraktaj nombroj de Lelong. Aktoj de Internacia Scienca Akademio Comenius, Beijing, vol. 1, pp. 56–65 (1992)
78. Klimek, M.: Extremal plurisubharmonic functions and invariant pseudodistances. Bull. Soc. Math. France **113**, 231–240 (1985)
79. Klimek, M.: Invariant pluricomplex Green functions. In: Topics in Complex Analysis, Warsaw, 1992. Banach Center Publications, vol. 31, pp. 207–226. Polish Academy of Sciences (1995)
80. Kobayashi, S.: Geometry of bounded domains. Trans. Am. Math. Soc. **92**, 267–290 (1959)

81. Kołodziej, S.: Some sufficient conditions for solvability of the Dirichlet problem for the complex Monge–Ampère operator. Ann. Pol. Math. **65**, 11–21 (1996)
82. Kołodziej, S.: The complex Monge–Ampère equation. Acta Math. **180**, 69–117 (1998)
83. Krylov, N.V.: Boundedly inhomogeneous elliptic and parabolic equations, Izv. Akad. Nauk SSSR **46**, 487–523 (1982) (English translation Math. USSR Izv. 20, 459–492, 1983)
84. Kuperberg, G.: From the Mahler conjecture to Gauss linking integrals. Geom. Funct. Anal. **18**, 870–892 (2008)
85. Lempert, L.: La métrique de Kobayashi et la représentation des domaines sur la boule. Bull. Soc. Math. France **109**, 427–474 (1981)
86. Lempert, L.: Private communication (2013)
87. Lu, Q.-K.: On Kaehler manifolds with constant curvature. Acta Math. Sin. **16**, 269–281 (1966)
88. Mahler, K.: Ein Minimalproblem für konvexe Polygone. Math. B (Zutphen) **7**, 118–127 (1938)
89. Maitani, F., Yamaguchi, H.: Variation of Bergman metrics on Riemann surfaces. Math. Ann. **330**, 477–489 (2004)
90. Martino, V., Montanari, A.: Integral formulas for a class of curvature PDE's and applications to isoperimetric inequalities and to symmetry problems. Forum Math. **22**, 255–267 (2010)
91. McNeal, J., Varolin, D.: Analytic inversion of adjunction: L^2 extension theorems with gain. Ann. Inst. Fourier **57**, 703–718 (2007)
92. Montanari, A., Lanconelli, E.: Pseudoconvex fully nonlinear partial differential operators: strong comparison theorems. J. Differ. Equ. **202**, 306–331 (2004)
93. Nazarov, F.: The Hörmander proof of the Bourgain–Milman theorem. In: Klartag, B., Mendelson, S., Milman, V.D. (eds) Geometric Aspects of Functional Analysis. Israel Seminar 2006–2010. Lecture Notes in Mathematics, vol 2050, pp. 335–343. Springer, New York (2012)
94. Nemirovskii, S.Yu., Shafikov, R.G.: Conjectures of Cheng and Ramadanov (Russian), Uspekhi Mat. Nauk **614**(370), 193–194 (2006) (translation in Russ. Math. Surv. 61, 780–782, 2006)
95. Nikolov, N., Pflug, P., Zwonek, W.: Estimates for invariant metrics on $\mathbb{C}$-convex domains. Trans. Am. Math. Soc. **363**, 6245–6256 (2011)
96. Ohsawa, T.: On the Bergman kernel of hyperconvex domains. Nagoya Math. J. **129**, 43–59 (1993)
97. Ohsawa, T.: Addendum to "On the Bergman kernel of hyperconvex domains". Nagoya Math. J. **137**, 145–148 (1995)
98. Ohsawa, T., Takegoshi, K.: On the extension of L^2 holomorphic functions. Math. Z. **195**, 197–204 (1987)
99. Pogorelov, A.V.: On the generalized solutions of the equation $\det(\partial^2 u/\partial x_i \partial x_j) = \varphi(x_1, x_2, \ldots, x_n) \geq 0$. Dokl. Akad. Nauk SSSR **200**, 534–537 (1971) (English translation Sov. Math. Dokl. **12**, 1436–1440, 1971)
100. Poletsky, E.A.: Holomorphic currents. Indiana Univ. Math. J. **42**, 85–144 (1993)
101. Reilly, R.C.: On the Hessian of a function and the curvatures of its graph. Mich. Math. J. **20**, 373–383 (1974)
102. Rothaus, O.S.: Some properties of Laplace transforms of measures. Trans. Am. Math. Soc. **131**, 163–169 (1968)
103. Ryabogin, D., Zvavitch, A.: Analytic methods in convex geometry. Lectures given at the Polish Academy of Sciences, 2011. In: IMPAN Lecture Notes. (2014, to appear)
104. Saint-Raymond, J.: Sur le volume des corps convexes symétriques. In: Initiation Seminar on Analysis: G. Choquet, M. Rogalski, J. Saint-Raymond, 20th Year: 1980/1981, Exp. No. 11, pp. 25. Publicationes Mathematicae, University Pierre et Marie Curie, vol. 46. University Paris VI, Paris (1981)
105. Santaló, L.A.: An affine invariant for convex bodies of n-dimensional space. Port. Math. **8**, 155–161 (1949), (in Spanish)
106. Schneider, R.: Convex Bodies: The Brunn–Minkowski Theory. Cambridge University Press, Cambridge (2014)
107. Siu, Y.T.: Analyticity of set sassociated to Lelong numbers and extension of closed positive currents. Invent. Math. **27**, 53–156 (1974)
108. Siu, Y.T.: Extension of meromorphic maps into Kähler manifolds. Ann. Math. **102**, 421–462 (1975)
109. Siu, Y.T.: The Fujita conjecture and the extension theorem of Ohsawa–Takegoshi. In: Geometric Complex Analysis, 1995, pp. 577–592. World Scientific, Hayama (1996)
110. Suita, N.: Capacities and kernels on Riemann surfaces. Arch. Ration. Mech. Anal. **46**, 212–217 (1972)
111. Talenti, G.: Elliptic equations and rearrangements. Ann. Scuola Norm. Sup. Pisa **3**, 697–718 (1976)

112. Talenti, G.: Some estimates of solutions to Monge–Ampère type equations in dimension two. Ann. Scuola Norm. Sup. Pisa **8**, 183–230 (1981)
113. Tso, K.: On symmetrization and Hessian equations. J. Anal. Math. **52**, 94–106 (1989)
114. Wiegerinck, J.: Domains with finite dimensional Bergman space. M. Zeit. **187**, 559–562 (1984)
115. Zakharyuta, V.P.: Spaces of analytic functions and maximal plurisubharmonic functions. D. Sc. Dissertation, Rostov-on-Don (1985)
116. Zeriahi, A.: Fonction de Green pluricomplexe à pôle à linfini sur un espace de Stein parabolique et applications. Math. Scand. **69**, 89–126 (1991)
117. Zwonek, W.: Regularity properties of the Azukawa metric. J. Math. Soc. Jpn. **52**, 899–914 (2000)
118. Zwonek, W.: An example concerning Bergman completeness. Nagoya Math. J. **164**, 89–102 (2001)
119. Zwonek, W.: Asymptotic behavior of the sectional curvature of the Bergman metric for annuli. Ann. Pol. Math. **98**, 291–299 (2010)

4

Seven pivotal theorems of Fourier analysis, signal analysis, numerical analysis and number theory: their interconnections

P. L. Butzer · M. M. Dodson · P. J. S. G. Ferreira · J. R. Higgins · G. Schmeisser · R. L. Stens

Abstract The present paper deals mainly with seven fundamental theorems of mathematical analysis, numerical analysis, and number theory, namely the generalized Parseval decomposition formula (GPDF), introduced 15 years ago, the well-known approximate sampling theorem (ASF), the new approximate reproducing kernel theorem, the basic Poisson summation formula, already known to Gauß, a newer version of the GPDF having a structure similar to that of the Poisson summation formula, namely,

In memory of Wolfgang Splettstößer (1950–2013), a pioneer in sampling analysis and valued colleague.

Communicated by S. K. Jain.

P. L. Butzer · R. L. Stens (✉)
Lehrstuhl A für Mathematik, RWTH Aachen University, 52056 Aachen, Germany
e-mail: stens@matha.rwth-aachen.de

P. L. Butzer
e-mail: butzer@rwth-aachen.de

M. M. Dodson
Department of Mathematics, University of York, York YO1O 5DD, UK
e-mail: mmd1@york.ac.uk

P. J. S. G. Ferreira
IEETA/DETI, Universidade de Aveiro, 3810-193 Aveiro, Portugal
e-mail: pjf@ua.pt

J. R. Higgins
I.H.P., 4 rue du Bary, 11250 Montclar, France
e-mail: rhiggins11@gmail.com

G. Schmeisser
Department of Mathematics, University of Erlangen-Nuremberg, 91058 Erlangen, Germany
e-mail: schmeisser@mi.uni-erlangen

the Parseval decomposition–Poisson summation formula, the functional equation of Riemann's zeta function, as well as the Euler–Maclaurin summation formula. It will in fact be shown that these seven theorems are all equivalent to one another, in the sense that each is a corollary of the others. Since these theorems can all be deduced from each other, one of them has to be proven independently in order to verify all. It is convenient to choose the ASF, introduced in 1963. The epilogue treats possible extensions to the more general contexts of reproducing kernel theory and of abstract harmonic analysis, using locally compact abelian groups. This paper is expository in the sense that it treats a number of mathematical theorems, their interconnections, their equivalence to one another. On the other hand, the proofs of the many intricate interconnections among these theorems are new in their essential steps and conclusions.

Keywords   Bandlimited and non-bandlimited functions · Sampling theorem · Parseval formula · Reproducing kernel formula · Poisson's summation formula · Riemann's zeta function · Euler–Maclaurin summation formula

Mathematics Subject Classfication   30D10 · 94A20 · 41A80 · 42A38 · 30D05

Contents

1 Introduction

A provoking result of Fourier analysis has turned up in the past 15 years, namely

1.1 Generalized Parseval decomposition formula (GPDF)

For $f \in F^2 \cap S_w^1$, $w > 0$, and $g \in F^2$, there holds $R_w f \in L^2(\mathbb{R})$ and

$$\int_{\mathbb{R}} f(u)\overline{g}(u)\, du = \frac{1}{w} \sum_{k \in \mathbb{Z}} f\left(\frac{k}{w}\right) \overline{g}\left(\frac{k}{w}\right)$$

$$- \frac{1}{w} \sum_{k \in \mathbb{Z}} f\left(\frac{k}{w}\right) \frac{1}{\sqrt{2\pi}} \int_{|v| \geq \pi w} \widehat{\overline{g}}(v) e^{ikv/w}\, dv + \int_{\mathbb{R}} (R_w f)(u)\overline{g}(u)\, du, \quad (1.1)$$

where

$$(R_w f)(t) := \frac{1}{\sqrt{2\pi}} \sum_{k \in \mathbb{Z}} \left(1 - e^{-i2\pi kwt}\right) \int_{(2k-1)\pi w}^{(2k+1)\pi w} \widehat{f}(v) e^{ivt}\, dv \quad (t \in \mathbb{R}). \quad (1.2)$$

Observe that $\lim_{w \to \infty} (R_w f)(t) = 0$ uniformly for $t \in \mathbb{R}$, since

$$\left| (R_w f)(t) \right| \leq \sqrt{\frac{2}{\pi}} \int_{|v| \geq \pi w} |\widehat{f}(u)|\, du. \quad (1.3)$$

Here F^2 and $F^2 \cap S_w^1$ are suitable subspaces of $L^2(\mathbb{R})$, and $\widehat{f}$ denotes the Fourier transform (for exact definitions see Sect. 2.1).

Formula (1.1), first established in [17, 18], can be said to be intermediate between the classical General Parseval/Plancherel formula (or power/energy theorem) for $L^2(\mathbb{R})$-functions, namely

$$\int_{\mathbb{R}} f(u)\overline{g}(u)\, du = \int_{\mathbb{R}} \widehat{f}(v)\overline{\widehat{g}}(v)\, dv \quad (f, g \in L^2(\mathbb{R})),$$

and the special case of (1.1) for bandlimited $L^2(\mathbb{R})$-functions, i.e., for functions $f, g \in \widehat{B}_{\pi w}^2$ (see Sect. 2.1), namely,

$$\int_{\mathbb{R}} f(u)\overline{g}(u)\, du = \frac{1}{w} \sum_{k \in \mathbb{Z}} f\left(\frac{k}{w}\right) \overline{g}\left(\frac{k}{w}\right) = \int_{\mathbb{R}} \widehat{f}(v)\overline{\widehat{g}}(v)\, dv \quad (f, g \in \widehat{B}_{\pi w}^2).$$

$$(1.4)$$

In fact, formula (1.1) is the extension of (1.4) from $\widehat{B}_{\pi w}^2$ to a larger subclass of $L^2(\mathbb{R})$.

Note that if $g \in \widehat{B}_{\pi w}^2$, then the hypotheses of GPDF are satisfied and the second series on the right-hand side of (1.1) vanishes, and so (1.1) reduces to

$$\int_{\mathbb{R}} f(u)\overline{g}(u)\,du = \frac{1}{w} \sum_{k\in\mathbb{Z}} f\left(\frac{k}{w}\right)\overline{g}\left(\frac{k}{w}\right) + \int_{\mathbb{R}} (R_w f)(u)\overline{g}(u)\,du. \qquad (1.5)$$

Moreover, if, in addition, $f \in \widehat{B}_{\pi w}^2$, then the integral on the right-hand side of (1.5) also vanishes since $(R_w f)(u) = 0$, and so (1.5) reduces to the (classical) General Parseval formula (GPF) for $f, g \in \widehat{B}_{\pi w}^2$, $w > 0$, already stated in (1.4). In this sense the latter two terms in GPDF (1.1) can be regarded as remainder terms.

In this instance the assumption $f \in S_w^1$ is implicitly contained in the hypothesis $f \in \widehat{B}_{\pi w}^2$; see Sect. 2.1.

The *second* formula to be considered is the approximate sampling formula (ASF) treated in Weiss [83], Brown [11,12] and Butzer-Splettstößer [22]; see also [84, pp. 64–66].

1.2 Approximate sampling formula (ASF)

For $f \in F^2 \cap S_w^1$, $w > 0$, we have

$$f(t) = \sum_{k\in\mathbb{Z}} f\left(\frac{k}{w}\right) \operatorname{sinc}(wt - k) + (R_w f)(t) \qquad (t \in \mathbb{R}), \qquad (1.6)$$

the series converging absolutely and uniformly on $\mathbb{R}$, where

$$\operatorname{sinc} t := \begin{cases} \dfrac{\sin \pi t}{\pi t}, & t \in \mathbb{R}\setminus\{0\}, \\ 1, & t = 0. \end{cases}$$

For the important role played by the sinc-function in mathematics see [80,81].

The ASF for not necessarily bandlimited functions generalizes the classical version

$$f(t) = \sum_{k\in\mathbb{Z}} f\left(\frac{k}{w}\right) \operatorname{sinc}(wt - k) \qquad (t \in \mathbb{R}) \qquad (1.7)$$

of the sampling formula for $f \in \widehat{B}_{\pi w}^2$.

In signal analysis, convergence the sampling series is usually defined via the limit of the symmetric partial sums, i.e., $\lim_{N\to\infty} \sum_{|k|\leq N} f(k/w) \operatorname{sinc}(wt - k)$ [see (2.20)], reflecting the mode of convergence of complex Fourier series (see, e.g., [41, pp. 5–6]). In particular, it is the usual approach in sampling theory, where the ASF holds for functions in F^2. In the present case, however, the focus is on the equivalence of the validity of the ASF with other results, in particular with the PSF (see Sect. 3.4).

Whereas to proceed from (1.7) to (1.6) one adds the "error" term $(R_w f)(t)$, to go from (1.4) to (1.1) one has to add two "error" terms.

To make the foregoing generalization fully clear, the ASF yields the GPDF and, conversely, the GPDF for just one special function will yield the ASF; thus the two formulae are truly equivalent to another, as already shown in [18].

A *third* formula belonging to our grouping of six equivalences in Fig. 1 brings us into contact with the theory of Hilbert spaces with reproducing kernel, an important branch of functional analysis and complex function theory having many areas of applications, with a long history. See, e. g., [3,37,46,62].

It is well known that $\widehat{B}_{\pi w}^2$ is a sub-Hilbert space of $L^2(\mathbb{R})$ and is, furthermore, an example of a Hilbert space with reproducing kernel. It is a special case of Saitoh's presentation (see, e. g., [41, Ch. 3], [77, pp. 9, 194]). The reproducing kernel is $(t, u) \mapsto w \operatorname{sinc} w(t-u)$ and the reproducing kernel formula (RKF), expressing the reproduction of $f \in \widehat{B}_{\pi w}^2$ from itself, that is, all of its values, is

$$f(t) = w \int_{\mathbb{R}} f(u) \operatorname{sinc} w(t - u) \, du = \int_{\mathbb{R}} f\left(\frac{\kappa}{w}\right) \operatorname{sinc}(wt - \kappa) \, d\kappa \quad (t \in \mathbb{R}). \quad (1.8)$$

The second integral in (1.8) is just the first integral rearranged to give a precise integral analogue of the series in (1.7), which may be thought of as a "discrete reproducing kernel formula", expressing the reproduction of f from a discrete subset of its values.

In view of the foregoing equivalence GPDF $\Leftrightarrow$ ASF, the question arises whether there can be established a generalized version of RKF, thus one for non-band-limited $L^1(\mathbb{R})$-functions, which is equivalent to the former two. In fact, one basic aim of this paper is to introduce the new approximate reproducing kernel formula, namely

1.3 Approximate reproducing kernel formula (ARKF)

Let $f \in F^2 \cap S_w^1$ with $w > 0$. Then $R_w f \in L^2(\mathbb{R})$ and

$$f(t) = w \int_{\mathbb{R}} f(u) \operatorname{sinc} w(t - u) \, du + (R_w f)(t) - w \int_{\mathbb{R}} (R_w f)(u) \operatorname{sinc} w(t - u) \, du.$$

$$(1.9)$$

Clearly, if $f \in \widehat{B}_{\pi w}^2$, then again $R_w f = 0$, and so we obtain the (classical) reproducing kernel formula RKF, namely (1.8). Thus (1.9) represents the classical RKF plus two remainder terms.

There is a *fourth* formula which is fundamental in a variety of mathematical fields, namely the truly well-known Poisson summation formula (PSF). But for purposes of precision let us state the version we are using.

1.4 Poisson's summation formula (PSF)

Let $f \in L^1(\mathbb{R})$ such that $\widehat{f} \in S_w^1$ for some $w > 0$; then

$$\sqrt{2\pi}\,w \sum_{k\in\mathbb{Z}} f(x + 2k\pi w) = \sum_{k\in\mathbb{Z}} \widehat{f}\left(\frac{k}{w}\right) e^{ikx/w} \quad (a.e.). \tag{1.10}$$

Now to the *fifth* formula which is actually a different version of GPDF but has a structure similar to that of the Poisson summation formula, however not for one function f, but for the product of two, namely $f\overline{g}$ on the left-hand side, and the product of their Fourier transforms on the right-hand side.

1.5 Parseval decomposition–Poisson summation formula (PDPS)

For $f \in F^2 \cap S_w^1$, $w > 0$, $\sigma > 0$, and $g \in F^2$, there holds

$$\frac{1}{w} \sum_{k\in\mathbb{Z}} f\left(\frac{k}{w}\right) \overline{g}\left(\frac{k}{w}\right) e^{ikx/w} = \sum_{k\in\mathbb{Z}} \int_{-\sigma}^{\sigma} \widehat{f}(v - x + 2k\pi w) \overline{\widehat{g}}(v)\, dv$$
$$+ \frac{1}{w} \sum_{k\in\mathbb{Z}} f\left(\frac{k}{w}\right) \frac{1}{\sqrt{2\pi}} \int_{|v|\geq\sigma} \overline{\widehat{g}}(v) e^{ik(v+x)/w}\, dv$$
$$\tag{1.11}$$

almost everywhere on $\mathbb{R}$. Furthermore, if rather $g \in F^1$ instead of $g \in F^2$, then the formula holds for all $x \in \mathbb{R}$.

PDPS can in fact be interpreted as an *approximate* form of PSF for a product $f\overline{g}$ in which g takes the role of an *approximately* bandlimited function. Indeed, the second series on the right-hand side of PDPS may be seen as an error term depending on the deviation of g from a bandlimited function. If g is bandlimited to $[-\sigma, \sigma]$, i.e. $g \in \widehat{B}_\sigma^2$, then PDPS reduces to

$$\frac{1}{w} \sum_{k\in\mathbb{Z}} f\left(\frac{k}{w}\right) \overline{g}\left(\frac{k}{w}\right) e^{ikx/w} = \sum_{k\in\mathbb{Z}} \int_{-\sigma}^{\sigma} \widehat{f}(v - x + 2k\pi w) \overline{\widehat{g}}(v)\, dv \tag{1.12}$$

holding *almost everywhere* on $\mathbb{R}$ when $g \in F^2$ and *everywhere* if rather $g \in F^1$.

A particular form of PDPS had already been used in [18] in one of the proofs, namely the form

$$\frac{1}{w} \sum_{k\in\mathbb{Z}} f\left(\frac{k}{w}\right) \overline{g}\left(\frac{k}{w}\right) = \sum_{k\in\mathbb{Z}} \int_{-\pi w}^{\pi w} \widehat{f}(v + 2k\pi w) \overline{\widehat{g}}(v)\, dv$$
$$+ \frac{1}{w} \sum_{k\in\mathbb{Z}} f\left(\frac{k}{w}\right) \frac{1}{\sqrt{2\pi}} \int_{|v|\geq\pi w} \overline{\widehat{g}}(v) e^{ikv/w}\, dv, \tag{1.13}$$

provided $f \in F^2 \cap S_w^1$ and $g \in F^1$. It is the case $x = 0$, $\sigma = \pi w$ of (1.11), noting

$$\int_{-\sigma}^{\sigma} \widehat{f}(v - x + 2k\pi w)\,\overline{\widehat{g}}(v)\,dv = \int_{2k\pi w - x - \sigma}^{2k\pi w - x + \sigma} \widehat{f}(v)\,\overline{\widehat{g}}(v + x - 2k\pi w)\,dv. \quad (1.14)$$

In fact, PDPS in the form (1.13) is just a different representation of GPDF; see [18] for the details.

Of interest is a version of (1.13) with g replaced by $\widehat{g}$, the Fourier transform of a function $g \in L^1(\mathbb{R}) \cap L^2(\mathbb{R})$, thus $\widehat{g} \in F^2$. The result can be regarded as a discretization of the classical Fourier exchange formula (2.5).

Just as for GPDF, the authors have also never met PDPS in the literature; both formulae also do not seem to be particular cases of specific trace formulae known to the authors.

Next to our *sixth* formula, the famous

1.6 Functional equation of the Riemann zeta function (FERZ)

For the Riemann zeta function, defined by

$$\zeta(s) := \sum_{k=1}^{\infty} \frac{1}{k^s} \quad (s = \sigma + i\tau \in \mathbb{C}, \sigma > 1), \quad (1.15)$$

and for $\sigma \leq 1$ by analytic continuation, there holds for all $s \in \mathbb{C}$ the equation

$$\pi^{-\frac{s}{2}} \Gamma\left(\frac{s}{2}\right) \zeta(s) = \pi^{-\frac{1-s}{2}} \Gamma\left(\frac{1-s}{2}\right) \zeta(1-s) \quad (s \in \mathbb{C})$$

or, equivalently,

$$2(2\pi)^{-s} \cos\left(\frac{\pi s}{2}\right) \Gamma(s)\zeta(s) = \zeta(1-s). \quad (1.16)$$

In fact, we shall reprove the well-known equivalence FERZ $\Leftrightarrow$ PSF (see the literature cited in Sect. 4) in great detail, in particular in regard to the interchanges of the orders of summation and/or integration occurring. What is remarkable about this equivalence is that in the direction FERZ $\Rightarrow$ PSF, a very particular result, namely the functional equation for $\zeta(s)$, implies a very general result, the PSF for a large class of functions. In other words, a formula of number theory, (externally) unrelated to periodicity and Fourier series, can yield a basic, general formula of analysis, one connecting Fourier series and integrals.

It is well-known that the FERZ is equivalent to the second famous functional equation, namely FEJT

$$\vartheta\left(-\frac{1}{z}\right) = (-iz)^{\frac{1}{2}} \vartheta(z), \quad (1.17)$$

satisfied by Jacobi's ϑ-function

$$\vartheta(z) = 1 + 2 \sum_{k=1}^{\infty} \exp(i\pi k^2 z) = \sum_{k=-\infty}^{\infty} \exp(i\pi k^2 z). \qquad (1.18)$$

The function ϑ, defined on the upper half-plane $H := \{z \in \mathbb{C}; \, \Im z > 0\}$, is analytic in H and periodic with period 2.

As to the proofs of FERZ $\Rightarrow$ FEJT see e.g. Hamburger [36, pp. 136–137], and for the converse Riemann [74]. In the matter see also [44].

It is well known that an application of PSF to the Gauß kernel of (2.12) yields FEJT (see [20, p. 204]).

Furthermore, Dedekind's η-function (1877) of elliptic modular functions,

$$\eta(z) = \exp\left(\frac{i\pi z}{12}\right) \prod_{k=1}^{\infty} \left(1 - \exp(i2\pi kz)\right), \qquad (1.19)$$

analytic on H, satisfies the same functional equation (1.17), namely FEDE (cf. [2, pp. 47–50]),

$$\eta\left(-\frac{1}{z}\right) = (-iz)^{\frac{1}{2}} \eta(z).$$

Finally to our *seventh* formula, discovered by Euler in connection with the so-called Basel problem, i.e., with determining $\zeta(2)$ in modern terminology.

1.7 Euler–Maclaurin summation formula (EMSF)

For $n, r \in \mathbb{N}$ and $f \in C^{(2r)}[0, n]$, we have

$$\sum_{k=0}^{n} f(k) = \int_0^n f(x)\, dx + \frac{1}{2}[f(0) + f(n)] + \sum_{k=1}^{r} \frac{B_{2k}}{(2k)!}\left[f^{(2k-1)}(n) - f^{(2k-1)}(0)\right]$$

$$+ (-1)^r \sum_{k=1}^{\infty} \int_0^n \frac{e^{i2\pi kt} + e^{-i2\pi kt}}{(2\pi k)^{2r}}\, f^{(2r)}(t)\, dt, \qquad (1.20)$$

where B_{2k} are the Bernoulli numbers.

The implications to be established in this paper are indicated by the arrows in Fig. 1. For a proof of ASF $\Leftrightarrow$ EMSF the reader is referred to [16]. Thus each of the seven formulae GPDF, ASF, ARKF, PSF, PDPS, FERZ and EMSF is equivalent to each other, in the sense that each is a corollary of each of the others.

This means that three different, particular results of number theory (FERZ, FEJT, FEDE) are equivalent to six general summation formulae of Fourier analysis, signal analysis and numerical analysis, namely PSF, PDPS, GPDF, ASF, ARKF, EMSF.

Section 2 is devoted to notations and the side results needed. In Sect. 3 we prove a first grouping of equivalences, namely, GPDF $\Leftrightarrow$ ASF, and then ASF $\Leftrightarrow$ ARKF, PSF $\Leftrightarrow$ PDPS, and finally ASF $\Leftrightarrow$ PSF. The latter three equivalences are new results

Fig. 1 The implications to be
proved

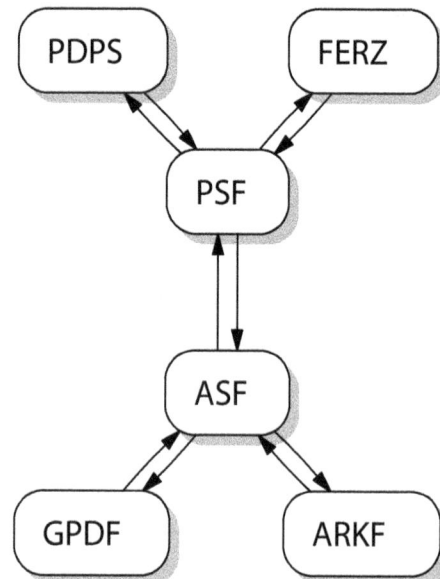

of this paper. Section 4 is concerned with the equivalence of PSF and FERZ, and in
Sect. 5 gives a proof of ASF independent of the other formulae of this paper.

The epilogue of Sect. 6, with the topic abstract settings, deals with the possibility
of presenting this paper in a setting based on reproducing kernel (r.k.) theory, and also
in terms of abstract harmonic analysis, using locally compact abelian groups.

In the present paper, dealing with the grouping (A): GPDF, ASF, ARKF, PSF,
PDPS, FERZ, as well as the Euler–Maclaurin summation formula EMSF (handled in
[16])—all for non-bandlimited functions—the basic result is that all seven theorems are
equivalent amongst themselves. On the other hand, in our joint paper [15], treating the
grouping (B), namely the classical sampling theorem CSF, the general Parseval formula
GPF (i. e., the GPDF for bandlimited functions), the reproducing kernel formula GPF
(i. e., GPDF for bandlimited functions) and the PSF—all for bandlimited functions—it
was shown that the latter four theorems are also all equivalent amongst themselves.

However in our article [14] it was shown that the approximate sampling theorem,
the ASF, is equivalent to the classical sampling theorem CSF. Thus the conclusion of
the present paper, together with the papers [15, 16] and [14], is the full equivalence of
the groupings (A) and (B). In other words, the seven theorems for non-bandlimited
functions are all equivalent to the corresponding ones for the classical bandlimited
functions.

Whereas the majority of the assertions of the theorems in question have been stated
and established in earlier papers by a (part) of the present authors, the proofs of the
many delicate interconnections between those of grouping (A), the chief topic of the
paper, are new in their essential steps and conclusions, as referred to.

Electrical engineers use bandlimited functions as a mathematical model which is a
rather severe restriction. In fact there do not exist signals which are simultaneously bandlimited and duration-limited (time-limited). Hence engineers work intuitively with

approximately bandlimited functions; these are covered by our non-bandlimited theorems. Thus our paper can also be regarded as a further justification of what electrical engineers do in their real life work.

2 Notations and side results

2.1 Notations

The Fourier transform of $f \in L^p(\mathbb{R})$ with $p = 1$ or $p = 2$ is defined by

$$\widehat{f}(v) := \frac{1}{\sqrt{2\pi}} \int_{\mathbb{R}} f(u) e^{-ivu} \, du \qquad (v \in \mathbb{R}),$$

the integral being understood as the limit in the $L^2(\mathbb{R})$-norm for $p = 2$. We shall consider the following function spaces, namely

$$F^p := \left\{ f \in L^p(\mathbb{R}) \cap C(\mathbb{R}) \, ; \, \widehat{f} \in L^1(\mathbb{R}) \right\} \qquad (p = 1, 2),$$

where $C(\mathbb{R})$ denotes the space of all uniformly continuous and bounded functions on $\mathbb{R}$, and

$$S_w^p := \left\{ f : \mathbb{R} \to \mathbb{C} \, ; \, \left(f(\tfrac{k}{w}) \right)_{k \in \mathbb{Z}} \in \ell^p(\mathbb{Z}) \right\} \qquad (w > 0, 1 \le p \le \infty).$$

There holds $F^1 \subset F^2$ and $S_w^1 \subset S_w^{p_1} \subset S_w^{p_2} \subset S_w^\infty$ for $1 \le p_1 \le p_2 \le \infty$.

The Bernstein spaces $\widehat{B}_\sigma^p$ for $p = 1, 2$ and $\sigma > 0$ are defined in terms of the Fourier transform via

$$\widehat{B}_\sigma^p := \left\{ f \in L^p(\mathbb{R}) \, ; \, \widehat{f}(v) = 0 \, a.\,e. \text{ outside } [-\sigma, \sigma] \right\}.$$

Since the Fourier transform of a function $f \in \widehat{B}_\sigma^p$ has support on a set of finite measure, those functions are also called "bandlimited". One has the inclusions $\widehat{B}_\sigma^1 \subset \widehat{B}_\sigma^2$ and $\widehat{B}_\sigma^p \subset F^p \cap S_w^1 \subset F^p \cap S_w^p$, $p = 1, 2$; see [64, pp. 123,126].

2.2 Results on Fourier analysis

Let us list some well-known results from Fourier analysis; see any textbook in the matter, e. g., [20, Chapt. 5.1 and 5.2].

Proposition 1 (a) *If $f \in L^p(\mathbb{R})$, $p = 1, 2$, then for each $h \in \mathbb{R}$,*

$$[f(\cdot + h)]\widehat{}(v) = \widehat{f}(v) e^{ihv} \tag{2.1}$$

$$[e^{-ih\cdot} f(\cdot)]\widehat{}(v) = \widehat{f}(v + h) \tag{2.2}$$

for all $v \in \mathbb{R}$ in case $p = 1$ and a. e. in case $p = 2$.

(b) *If $f \in L^1(\mathbb{R})$, $g \in L^p(\mathbb{R})$, $p = 1, 2$, then the convolution*

$$(f * g)(t) := \frac{1}{\sqrt{2\pi}} \int_{\mathbb{R}} f(u)g(t - u)\, du \qquad (2.3)$$

belongs to $L^p(\mathbb{R})$, and for the Fourier transform one has the convolution theorem

$$(f * g)\widehat{}(v) = \widehat{f}(v)\widehat{g}(v) \qquad (2.4)$$

for all $v \in \mathbb{R}$ in case $p = 1$ and a. e. in case $p = 2$.

(c) *If f, $g \in L^p(\mathbb{R})$, $p = 1, 2$, then there holds the exchange formula*

$$\int_{\mathbb{R}} f(v)\widehat{g}(v)\, dv = \int_{\mathbb{R}} \widehat{f}(v)g(v)\, dv. \qquad (2.5)$$

(d) *If $f \in L^2(\mathbb{R})$, then $\widehat{f}$ also belongs to $L^2(\mathbb{R})$ and $\|f\|_{L^2(\mathbb{R})} = \|\widehat{f}\|_{L^2(\mathbb{R})}$. Furthermore, there holds the inversion formula*

$$f(t) = \int_{\mathbb{R}} \widehat{f}(v)e^{ivt}\, dv = \widehat{\widehat{f}}(-t) \qquad a.\,e., \qquad (2.6)$$

where the integral is again understood as the limit in $L^2(\mathbb{R})$-norm. If $f \in F^p$ with $p = 1$ or $p = 2$, then the integral in (2.6) exists as an ordinary Lebesgue integral and both equalities hold for all $t \in \mathbb{R}$.

Lemma 1 *There hold the formulae*

$$\frac{1}{\sqrt{2\pi}\, w} \operatorname{rect}\left(\frac{\cdot}{w}\right)\widehat{}(v) = \operatorname{sinc}(wv) \qquad (v \in \mathbb{R}), \qquad (2.7)$$

$$\operatorname{sinc}(w\cdot)\widehat{}(v) = \frac{1}{\sqrt{2\pi}\, w} \operatorname{rect}\left(\frac{v}{w}\right) \qquad (v \in \mathbb{R}) \qquad (2.8)$$

with the rectangle function

$$\operatorname{rect}(t) := \begin{cases} 1, & |t| < \pi, \\ \frac{1}{2}, & |t| = \pi, \\ 0, & |t| > \pi. \end{cases}$$

Furthermore, one has

$$\int_{\mathbb{R}} \operatorname{sinc}(u - s)\, \operatorname{sinc}(u - t)\, du = \operatorname{sinc}(s - t) \qquad (s, t \in \mathbb{R}), \qquad (2.9)$$

and, in particular,

$$\int_{\mathbb{R}} \operatorname{sinc}(u - j)\, \operatorname{sinc}(u - k)\, du = \begin{cases} 1, & j = k, \\ 0, & j \neq k. \end{cases} \qquad (j, k \in \mathbb{Z}). \qquad (2.10)$$

Proof Equation (2.7) follows by a simple integration, and (2.8) follows from (2.7) by the Fourier inversion formula. (For an elementary proof of (2.8) without using the inversion formula see [10, p. 13 f].) As to (2.9) one has by Proposition 1 (a), (c), and by (2.7) and (2.8)

$$
\begin{aligned}
\int_{\mathbb{R}} \operatorname{sinc}(u - s) \operatorname{sinc}(u - t) \, du &= \int_{\mathbb{R}} \left[\frac{1}{\sqrt{2\pi}} \operatorname{rect}(\cdot) e^{i \cdot s} \right]^{\wedge} (u) \operatorname{sinc}(u - t) \, du \\
&= \int_{\mathbb{R}} \left[\frac{1}{\sqrt{2\pi}} \operatorname{rect}(v) e^{ivs} \right] \operatorname{sinc}(\cdot - t)^{\wedge}(v) \, du \\
&= \int_{\mathbb{R}} \left[\frac{1}{\sqrt{2\pi}} \operatorname{rect}(v) e^{ivs} \right] \left[\frac{1}{\sqrt{2\pi}} \operatorname{rect}(v) e^{-ivt} \right] dv \\
&= \frac{1}{2\pi} \int_{\mathbb{R}} \operatorname{rect}(v) e^{-iv(t-s)} \, dv = \frac{1}{\sqrt{2\pi}} \operatorname{rect}^{\wedge}(t - s) \\
&= \operatorname{sinc}(t - s) = \operatorname{sinc}(s - t).
\end{aligned}
$$

Finally, (2.10) follows from (2.9) noting that $\operatorname{sinc}(j - k) = \delta_{j,k}$. $\qquad\square$

We shall also need the following result for the approximate identities g_ρ of Gauß-Weierstraß and χ_ρ of Fejér; see [20, Sec. 3.1.3, 3.1.2, 3.2].

Proposition 2 (a) *For $f \in C(\mathbb{R})$, we have*

$$
\lim_{\rho \to 0+} \frac{1}{\sqrt{2\pi}} \int_{\mathbb{R}} f(u) g_\rho(t - u) \, du = f(t) \tag{2.11}
$$

uniformly for $t \in \mathbb{R}$, where

$$
g_\rho(u) := \frac{1}{\sqrt{2\rho}} \exp\left(\frac{-u^2}{4\rho} \right) \quad (u \in \mathbb{R}; \rho > 0) \tag{2.12}
$$

is the Gaussian kernel, its Fourier transform being

$$
\widehat{g}_\rho(v) = e^{-\rho v^2} \quad (v \in \mathbb{R}; \rho > 0). \tag{2.13}
$$

The same result is valid for $f \in L^p(\mathbb{R})$, $1 \le p < \infty$; the convergence in (2.11) holding now in $L^p(\mathbb{R})$-norm as well as a. e. on $\mathbb{R}$.

(b) *The assertions of part (a) remain valid, if the Gaussian kernel is replaced by Fejér's kernel*

$$
\chi_\rho(u) := \rho \sqrt{2\pi} \operatorname{sinc}^2(\rho u) \quad (u \in \mathbb{R}, \rho > 0), \tag{2.14}
$$

having Fourier transform

$$
\widehat{\chi}_\rho(v) = \left(1 - \frac{|v|}{2\pi\rho} \right)_+ \quad (v \in \mathbb{R}, \rho > 0), \tag{2.15}
$$

where $h_+(u) := h(u)$ if $h(u) \ge 0$, and $h_+(u) = 0$ if $h(u) < 0$.

For $f := \text{sinc}(\cdot - u)$, we obtain from ASF the following sinc summation formula:

$$\text{sinc}(t - u) = \sum_{k \in \mathbb{Z}} \text{sinc}(t - k)\,\text{sinc}(u - k). \tag{2.16}$$

However, we need not invoke ASF for deriving (2.16). It is known that (2.16) is a simple consequence of the famous, classical cotangent expansion. For the latter, easy elementary proofs are known; see e. g. [43], [15, p. 448]

It follows from (2.16) that

$$\sum_{k \in \mathbb{Z}} \text{sinc}^2(t - k) = 1 \qquad (t \in \mathbb{R}). \tag{2.17}$$

The following lemma will be useful; for a proof see [40, p. 207].

Lemma 2 *Let $f \in L^p(\mathbb{R})$, $1 < p < \infty$, and $(f_k)_{k \in \mathbb{N}}$ be a sequence in $L^p(\mathbb{R})$. If there exists a constant M such that $\|f_k\|_{L^p(\mathbb{R})} \leq M$ for all $k \in \mathbb{N}$, and if $\lim_{k \to \infty} f_k(u) = f(u)$ a. e. in $\mathbb{R}$, then for every $g \in L^q(\mathbb{R})$, $1/p + 1/q = 1$,*

$$\lim_{k \to \infty} \int_{\mathbb{R}} f_k(u)g(u)\,du = \int_{\mathbb{R}} f(u)g(u)\,du.$$

2.3 The sampling series $S_w f$

Concerning the convergence of the series

$$(S_w f)(t) := \sum_{k \in \mathbb{Z}} f\left(\frac{k}{w}\right) \text{sinc}(wt - k) \qquad (t \in \mathbb{R}), \tag{2.18}$$

occurring in (1.6), (1.7), we have

Lemma 3 *Let $f \in S_w^1$ for some $w > 0$. The series $(S_w f)(t)$ converges absolutely and uniformly on $\mathbb{R}$. Furthermore, $S_w f$ converges in $L^p(\mathbb{R})$, $1 < p < \infty$. In particular, $S_w f \in L^p(\mathbb{R})$, and*

$$\lim_{n \to \infty} \int_{\mathbb{R}} (S_{w,n} f)(u)\overline{g}(u)\,du = \int_{\mathbb{R}} (S_w f)(u)\overline{g}(u)\,du \tag{2.19}$$

for all $g \in L^q(\mathbb{R})$, $1/p + 1/q = 1$, where

$$(S_{w,n} f)(t) := \sum_{|k| \leq n} f\left(\frac{k}{w}\right) \text{sinc}(wt - k) \tag{2.20}$$

denote the partial sums of $S_w f$.

Proof For simplicity take $w = 1$. For $0 < n_1 < n_2$ there holds

$$\left\| S_{1,n_1} f - S_{1,n_1} f \right\|_{L^p(\mathbb{R})} \leq \sum_{n_1 < |k| \leq n_2} |f(k)| \, \| \operatorname{sinc}(\cdot - k) \|_{L^p(\mathbb{R})}$$

$$= \| \operatorname{sinc} \|_{L^p(\mathbb{R})} \sum_{n_1 < |k| \leq n_2} |f(k)|, \qquad (2.21)$$

where the latter term tends to zero for $n_1, n_2 \to \infty$ since $f \in S_1^1$. The same estimate holds for the space $C(\mathbb{R})$, giving the absolute and uniform convergence of $(S_{1,n} f)(t)$ to $(S_1 f)(t)$. Further, the completeness of $L^p(\mathbb{R})$ and estimate (2.21) guarantee that $S_{1,n} f$ tends to some function $g \in L^p(\mathbb{R})$ which must be equal to the uniform limit $S_1 f$.

Finally, by Hölder's inequality

$$\left| \int_{\mathbb{R}} (S_{w,n} f)(u) \overline{g}(u) \, du - \int_{\mathbb{R}} (S_w f)(u) \overline{g}(u) \, du \right| \leq \| S_{w,n} f - S_w f \|_{L^p(\mathbb{R})} \| g \|_{L^q(\mathbb{R})},$$

which yields (2.19). □

2.4 Remarks on Poisson's summation formula

As to the PSF, let $h(x) := \sqrt{2\pi} w \sum_{k \in \mathbb{Z}} |f(x + 2k\pi w)|$; then

$$\int_{-\pi w}^{\pi w} h(x) \, dx = \sqrt{2\pi} w \sum_{k \in \mathbb{Z}} \int_{-\pi w}^{\pi w} |f(x + 2k\pi w)| \, dx$$

$$= \sqrt{2\pi} w \sum_{k \in \mathbb{Z}} \int_{(2k-1)\pi w}^{(2k+1)\pi w} |f(x)| \, dx = \sqrt{2\pi} w \int_{\mathbb{R}} |f(x)| \, dx < \infty, \qquad (2.22)$$

showing that h is integrable over the interval $[-\pi w, \pi w]$, and hence over any compact interval in view of the $2\pi w$-periodicity of h.

Now one has for arbitrary $a, b \in \mathbb{R}$ that

$$\int_a^b \left| \sqrt{2\pi} w \sum_{k=m}^{n} f(x + 2k\pi w) \right| dx \leq \int_a^b h(x) \, dx < \infty \quad (m, n \in \mathbb{Z}), \quad (2.23)$$

and so it follows that the left-hand side of (1.10) converges a. e. on $\mathbb{R}$, and the convergence is dominated by the locally integrable function h.

When we wish to deduce PSF from any of the other formulae, it suffices to prove it for a dense subspace of $X \subset L^1(\mathbb{R})$ with $\widehat{g} \in S_w^1$ for all $g \in X$. Indeed, let $f \in L^1(\mathbb{R})$ with $\widehat{f} \in S_w^1$, and let $(f_n)_{n \in \mathbb{N}}$ be a sequence in X with $\lim_{n \to \infty} \| f_n - f \|_{L^1(\mathbb{R})} = 0$. Denoting $f^*(x) := \sqrt{2\pi} w \sum_{k \in \mathbb{Z}} f(x + 2k\pi w)$, $f_n^*(x) := \sqrt{2\pi} w \sum_{k \in \mathbb{Z}} f_n(x + 2k\pi w)$, then (cf. (2.23), (2.22)),

$$\int_{-\pi w}^{\pi w} |f_n^*(u) - f^*(u)|du \leq \|f_n - f\|_{L^1(\mathbb{R})} \quad (n \in \mathbb{N}). \tag{2.24}$$

This yields that

$$\lim_{n \to \infty} \widehat{f_n^*}(k) = \widehat{f^*}(k) \quad (k \in \mathbb{Z}), \tag{2.25}$$

where $\widehat{f_n^*}(k)$ and $\widehat{f^*}(k)$ denote the trigonometric Fourier coefficients of the $2\pi w$-periodic function f_n^* and f^*, respectively.

On the other hand, PSF applied to the f_n states that

$$f_n^*(x) = \sum_{k \in \mathbb{Z}} \widehat{f_n}\left(\frac{k}{w}\right) e^{ikx/w} \quad a.e.,$$

meaning, in view of the uniform convergence of the series on the right, that $\widehat{f_n}(k/w)$, $k \in \mathbb{Z}$, are the Fourier coefficients of the $2\pi w$-periodic function on the left. This implies that

$$\widehat{f^*}(k) = \lim_{n \to \infty} \widehat{f_n^*}(k) = \lim_{n \to \infty} \widehat{f_n}\left(\frac{k}{w}\right) = \widehat{f}\left(\frac{k}{w}\right) \quad ((k \in \mathbb{Z}), \tag{2.26}$$

the last equality being valid in view of $|\widehat{f_n}(v) - \widehat{f}(v)| \leq \|f_n - f\|_{L^1(\mathbb{R})}$ for all $v \in \mathbb{R}$. Hence the Fourier coefficients of f^* are given by $\widehat{f^*}(k) = \widehat{f}(k/w)$, $k \in \mathbb{Z}$, i.e.

$$f^*(x) = \sum_{k \in \mathbb{Z}} \widehat{f}\left(\frac{k}{w}\right) e^{ikx/w} \quad a.e.$$

The equality a.e. here holds in view of $\widehat{f} \in S_w^1$, giving the uniform convergence of the Fourier series of f^*. This is PSF for all $f \in L^1(\mathbb{R})$ with $\widehat{f} \in S_w^1$.

3 Equivalence of the decomposition formula and four summation formulae

In this section we shall deduce the basic new ARKF theorem for $L^2(\mathbb{R})$-function by means of the ASF. Even more so, the ASF itself is a consequence of ARKF. Thus the two are equivalent. On top, three other basic equivalences of this paper are also established in this section.

When we want to deduce one of the formulae GPDF, ASF or ARKF, it suffices to carry it out for $w = 1$. Indeed, the general formula is obtained from that special case by replacing f by $f(\cdot/w)$, t by wt and noting that

$$\left[(R_1 f)\left(\frac{\cdot}{w}\right)\right](wt) = (R_w f)(t).$$

3.1 GPDF and ASF

But first to the equivalence of GPDF with ASF. The proof of the implication GPDF $\Rightarrow$ ASF is essentially that of Thm 1.3 in [18]. We present it here again for convenience.

Proof of GPDF $\Rightarrow$ ASF for $f \in F^2 \cap S_w^1$. In regard to the proof which follows, it is appropriate to rewrite (1.1) as

$$\int_{\mathbb{R}} f(u)\overline{g}(u)\,du = \frac{1}{w}\sum_{k\in\mathbb{Z}} f\left(\frac{k}{w}\right)\frac{1}{\sqrt{2\pi}}\int_{|v|\leq\pi w} \widehat{\overline{g}}(v)\,e^{ikv/w}\,dv$$
$$+ \int_{\mathbb{R}} (R_w f)(u)\overline{g}(u)\,du. \qquad (3.1)$$

Indeed, one has in view of the Fourier inversion formula (2.6) for g

$$\frac{1}{w}\sum_{k\in\mathbb{Z}} f\left(\frac{k}{w}\right)\overline{g}\left(\frac{k}{w}\right) - \frac{1}{w}\sum_{k\in\mathbb{Z}} f\left(\frac{k}{w}\right)\frac{1}{\sqrt{2\pi}}\int_{|v|\geq\pi w} \widehat{\overline{g}}(v)e^{ikv/w}\,dv$$
$$= \frac{1}{w}\sum_{k\in\mathbb{Z}} f\left(\frac{k}{w}\right)\frac{1}{\sqrt{2\pi}}\left\{\int_{-\infty}^{\infty} \widehat{\overline{g}}(v)e^{ikv/w}\,dv - \int_{|v|\geq\pi w} \widehat{\overline{g}}(v)e^{ikv/w}\,dv\right\}.$$

A substitution of this identity into (1.1) yields (3.1).

Now we restrict the matter to $w = 1$ and apply the GPDF in the form (3.1) to $f \in F^2 \cap S_1^1$ and $g = (2\pi)^{-1/2}g_\rho(t - \cdot) \in F^2$ of (2.12). Noting that $\widehat{g_\rho(t - \cdot)}(v) = e^{-\rho v^2}e^{-itv}$ by (2.1) and (2.13), we obtain

$$\frac{1}{\sqrt{2\pi}}\int_{\mathbb{R}} f(u)g_\rho(t - u)\,du = \sum_{k\in\mathbb{Z}} f(k)\frac{1}{2\pi}\int_{|v|\leq\pi} e^{-\rho v^2}e^{iv(k-t)}\,dv$$
$$+ \frac{1}{\sqrt{2\pi}}\int_{\mathbb{R}} (R_1 f)(u)g_\rho(t - u)\,du.$$

In this equation, we let $\rho \to 0+$. Since f and $R_1 f$ belong to $C(\mathbb{R})$, and since the infinite series is uniformly convergent in view of $f \in S_1^1$, Proposition 2 yields

$$f(t) = \sum_{k\in\mathbb{Z}} f(k)\frac{1}{2\pi}\int_{|v|\leq\pi} e^{iv(k-t)}\,dv + (R_1 f)(t) \qquad (t \in \mathbb{R}).$$

This is ASF, because the integral equals $2\pi\,\mathrm{sinc}(t - k)$. The absolute and uniform convergence of the series $(S_1 f)(t)$ follows from Lemma 3. □

The proof of the converse direction ASF $\Rightarrow$ GPDF is similar to the corresponding one in [18] for non-uniform sampling.

Proof of ASF $\Rightarrow$ GPDF for $f \in F^2 \cap S_w^1$, $g \in F^2$. Let $w = 1$, multiply ASF by $\overline{g}$ and integrate over $\mathbb{R}$. Then

$$\int_{\mathbb{R}} f(u)\overline{g}(u)\,du = \int_{\mathbb{R}} (S_1 f)(u)\overline{g}(u)\,du + \int_{\mathbb{R}} (R_1 f)(u)\overline{g}(u)\,du. \qquad (3.2)$$

Since $f \in L^2(\mathbb{R})$ by assumption, we also have $S_1 f \in L^2(\mathbb{R})$ in view of Lemma 3, and hence $R_1 f = f - S_w f \in L^2(\mathbb{R})$. Thus all integrals in (3.2) exists. For the first term on the right-hand side of (3.2) one obtains

$$\int_{\mathbb{R}} (S_1 f)(u)\overline{g}(u)\,du = \sum_{k \in \mathbb{Z}} f(k) \int_{\mathbb{R}} \overline{g}(u)\,\mathrm{sinc}(u - k)\,du, \qquad (3.3)$$

the interchange of integration and summation being justified, since $\big(f(k)\big) \in l^1(\mathbb{Z})$ and the integrals on the right-hand side are bounded by $\|g\|_{L^2(\mathbb{R})} \cdot \|\,\mathrm{sinc}\,\|_{L^2(\mathbb{R})}$. Now, using the convolution theorem (2.4), formula (2.8), and the Fourier inversion formula (2.6), we can rewrite the latter integral as

$$\int_{\mathbb{R}} \overline{g}(u)\,\mathrm{sinc}(u - k)\,du = \sqrt{2\pi}\,(\overline{g} * \mathrm{sinc})(k) = \frac{1}{\sqrt{2\pi}} \int_{\mathbb{R}} \widehat{\overline{g}}(v)\,\mathrm{rect}(v)e^{ikv}\,dv$$

$$= \frac{1}{\sqrt{2\pi}} \left\{ \int_{\mathbb{R}} - \int_{|v| \geq \pi} \right\} \widehat{\overline{g}}(v)e^{ikv}\,dv = \overline{g}(k) - \frac{1}{\sqrt{2\pi}} \int_{|v| \geq \pi} \widehat{\overline{g}}(v)e^{ikv}\,dv$$

Inserting this into (3.3), and then the resulting equation into (3.2) yields the assertion (1.1) for $w = 1$. □

Thus we have established the equivalence ASF $\Leftrightarrow$ GPDF.

3.2 ASF and ARKF

Now to the basic equivalence ASF $\Leftrightarrow$ ARKF mentioned above.

Proof of ASF $\Rightarrow$ ARKF for $f \in F^2 \cap S_w^1$, $g \in F^2$. The fact that $R_w f = f - S_w f \in L^2(\mathbb{R})$ follows again from Lemma 3. Now, let $w = 1$ and let f be as in ARKF. We consider ASF (1.6), replace t by u, multiply both sides by $\mathrm{sinc}(t - u)$ and integrate with respect to u. This gives

$$\int_{\mathbb{R}} f(u)\,\mathrm{sinc}(t - u)\,du = \int_{\mathbb{R}} \sum_{k \in \mathbb{Z}} f(k)\,\mathrm{sinc}(u - k)\,\mathrm{sinc}(t - u)\,du$$

$$+ \int_{\mathbb{R}} (R_1 f)(u)\,\mathrm{sinc}(t - u)\,du. \qquad (3.4)$$

By Lemma 3 we can interchange integration and summation on the right-hand side and we obtain by (2.9)

$$\int_{\mathbb{R}} \sum_{k \in \mathbb{Z}} f(k) \operatorname{sinc}(u - k) \operatorname{sinc}(t - u) \, du$$

$$= \sum_{k \in \mathbb{Z}} f(k) \int_{\mathbb{R}} \operatorname{sinc}(u - k) \operatorname{sinc}(t - u) \, du = \sum_{k \in \mathbb{Z}} f(k) \operatorname{sinc}(t - k).$$

Applying ASF once more, the latter series can be replaced by $f(t) - (R_1 f)(t)$, and hence

$$\int_{\mathbb{R}} \sum_{k \in \mathbb{Z}} f(k) \operatorname{sinc}(u - k) \operatorname{sinc}(t - u) \, du = f(t) - (R_1 f)(t).$$

Inserting this into (3.4) yields the ARKF in the form (1.9).   □

Proof of ARKF $\Rightarrow$ ASF for $f \in F^2 \cap S_1^1$. It suffices to show that

$$\int_{\mathbb{R}} [f(u) - (R_1 f)(u)] \operatorname{sinc}(t - u) \, du = \sum_{k \in \mathbb{Z}} f(k) \operatorname{sinc}(t - k), \qquad (3.5)$$

since the integral equals $f(t) - (R_1 f)(t)$. For this purpose, we shall use the sinc summation formula (2.16) on the left-hand side and interchange integration and summation. For justifying the interchange, we want to employ Lemma 2. We have to verify that the hypotheses of that result are satisfied in the present situation.

Clearly, $f - R_1 f$ and $\operatorname{sinc}(\cdot - t)$ belong to $L^2(\mathbb{R})$. Next, setting

$$\operatorname{sinc}_N(t - u) := \sum_{|k| \leq N} \operatorname{sinc}(u - k) \operatorname{sinc}(t - k),$$

we have $\lim_{N \to \infty} \operatorname{sinc}_N(t - u) = \operatorname{sinc}(t - u)$. Furthermore, by employing (2.9) and (2.17), we find that

$$\| \operatorname{sinc}_N(t - \cdot) \|_{L^2(\mathbb{R})}^2$$

$$= \sum_{|k| \leq N} \sum_{|l| \leq N} \left\{ \int_{\mathbb{R}} \operatorname{sinc}(u - k) \operatorname{sinc}(u - l) \, du \right\} \operatorname{sinc}(t - k) \operatorname{sinc}(t - l)$$

$$= \sum_{|k| \leq N} \sum_{|l| \leq N} \delta_{k,l} \cdot \operatorname{sinc}(t - k) \operatorname{sinc}(t - l)$$

$$= \sum_{|k| \leq N} \operatorname{sinc}^2(t - k) \leq \sum_{k \in \mathbb{Z}} \operatorname{sinc}^2(t - k) = 1,$$

where Kronecker's delta has been used. Now Lemma 2 allows us to conclude that

$$\int_{\mathbb{R}} [f(u) - R_1 f(u)] \operatorname{sinc}(t - u) \, du$$

$$= \lim_{N \to \infty} \int_{\mathbb{R}} [f(u) - R_1 f(u)] \operatorname{sinc}_N(t - u) \, du$$

$$= \lim_{N \to \infty} \sum_{|k| \leq N} \left\{ \int_{\mathbb{R}} [f(u) - R_1 f(u)] \operatorname{sinc}(u - k) \, du \right\} \operatorname{sinc}(t - k).$$

By ARKF for $t = k$, the expression in braces is equal to $f(k) - R_1 f(k)$. Since $R_1 f$ vanishes at the integers, we see that (3.5) holds. The statement regarding the convergence of the series is verified as in the proof of GPDF $\Rightarrow$ ASF.   □

The equivalence assertion ASF $\Leftrightarrow$ ARKF, just established, is another new result of this paper.

3.3 PSF and PDPS

The following concerns the final grouping of this section.

Proof of PSF $\Rightarrow$ PDPS It suffices again to consider the case $w = 1$. First we assume that $g \in F^2$ only and define

$$\overline{g}_1(u) := \frac{1}{\sqrt{2\pi}} \int_{-\sigma}^{\sigma} \widehat{\overline{g}}(v) e^{iuv} \, dv \quad \text{and} \quad \phi(u) := \left(f \overline{g}_1 \right)\widehat{\ }(-u),$$

which implies that $\widehat{\phi} = f \overline{g}_1$. Then $\widehat{\phi} \in L^1(\mathbb{R})$ since it is a product of two functions from $L^2(\mathbb{R})$, and $\widehat{\phi} \in S_1^1$ since $f \in S_1^1$ and $\overline{g}_1$ is bounded. Furthermore, $\phi \in C(\mathbb{R})$ since $\phi(-\cdot)$ is the Fourier transform of a function from $L^1(\mathbb{R})$. Now one has by Proposition 1 (a), (c) that

$$\begin{aligned}
\phi(-\xi) = \left(f \overline{g}_1 \right) \widehat{f}(\xi) &= \frac{1}{\sqrt{2\pi}} \int_{\mathbb{R}} f(v) \overline{g}_1(v) e^{-iv\xi} \, dv \\
&= \frac{1}{\sqrt{2\pi}} \int_{\mathbb{R}} \left[f(\cdot) e^{-i\xi \cdot} \right] \widehat{f}(-v) \overline{g}_1(v) \, dv \\
&= \frac{1}{\sqrt{2\pi}} \int_{\mathbb{R}} \widehat{f}(\xi - v) \widehat{\overline{g}}_1(v) \, dv.
\end{aligned} \tag{3.6}$$

Hence $\phi(-\cdot)$ is the convolution of $\widehat{f} \in L^1(\mathbb{R})$ and $\widehat{\overline{g}}_1 \in L^1(\mathbb{R})$, and so it follows by Proposition 1 b) that $\phi \in L^1(\mathbb{R})$. Altogether we have shown that ϕ satisfies the hypotheses of PSF for $w = 1$, and so

$$\sqrt{2\pi} \sum_{k \in \mathbb{Z}} \phi(x + 2k\pi) = \sum_{k \in \mathbb{Z}} \widehat{\phi}(k) e^{ikx} = \sum_{k \in \mathbb{Z}} f(k) \overline{g}_1(k) e^{ikx} \quad a.e. \tag{3.7}$$

Employing (3.6) and noting that $\widehat{\overline{g}}_1(v) = \overline{\widehat{g}}_1(-v) = \overline{\widehat{g}}(-v) \operatorname{rect}(v\pi/\sigma)$, we find that

$$\phi(\xi) = \left(f \overline{g}_1 \right) \widehat{f}(-\xi) = \frac{1}{\sqrt{2\pi}} \int_{-\sigma}^{\sigma} \widehat{f}(v - \xi) \overline{\widehat{g}}(v) \, dv.$$

Substituting this into the left-hand side of (3.7) yields

$$\sum_{k \in \mathbb{Z}} \int_{-\sigma}^{\sigma} \widehat{f}(v - x - 2k\pi) \overline{\widehat{g}}(v) \, dv = \sum_{k \in \mathbb{Z}} f(k) \overline{g}_1(k) e^{ikx} \quad a.e. \tag{3.8}$$

Using the Fourier inversion formula to obtain

$$\overline{g}_1(u) = \overline{g}(u) - \frac{1}{\sqrt{2\pi}} \int_{|v| \geq \sigma} \widehat{\overline{\overline{g}}}(v) e^{iuv} \, dv$$

for eliminating $\overline{g}_1$ on the right-hand side of (3.8), we arrive at

$$\sum_{k \in \mathbb{Z}} \int_{-\sigma}^{\sigma} \widehat{f}(v - x - 2k\pi) \, \overline{\overline{g}}(v) \, dv$$

$$= \sum_{k \in \mathbb{Z}} f(k) \, \overline{g}(k) e^{ikx} - \sum_{k \in \mathbb{Z}} f(k) \frac{1}{\sqrt{2\pi}} \int_{|v| \geq \sigma} \widehat{\overline{\overline{g}}}(v) e^{ik(v+x)} \, dv \quad a.e. \quad (3.9)$$

The desired formula is obtained by replacing k by $-k$ on the left-hand side.

When $g \in F^1$, then $\overline{\overline{g}} \in C(\mathbb{R})$. Since also $f \in S_1^1$ and $\overline{\overline{g}} \in L^1(\mathbb{R})$, we observe that the two series on the right-hand side of (3.9) depend continuously on x.

As to the left-hand side, choose $n \in \mathbb{N}$ such that $n\pi \geq \sigma$, then,

$$\left| \sum_{k \in \mathbb{Z}} \int_{-\sigma}^{\sigma} \widehat{f}(v - 2k\pi) \, \overline{\overline{g}}(v) dv \right| \leq \|\widehat{g}\|_{L^\infty(\mathbb{R})} \sum_{k \in \mathbb{Z}} \int_{-n\pi}^{n\pi} \left| \widehat{f}(v - 2k\pi) \right| dv$$

$$= \|\widehat{g}\|_{L^\infty(\mathbb{R})} \int_{-n\pi}^{n\pi} \sum_{k \in \mathbb{Z}} \left| \widehat{f}(v - 2k\pi) \right| dv. \quad (3.10)$$

Here and in the next displayed formula the interchange of summation and integration is justified by Beppo Levi's theorem.

Since the integrand in the last line of (3.10) is a 2π-periodic function, the integral equals n-times the integral from $-\pi$ to π, and hence

$$\left| \sum_{k \in \mathbb{Z}} \int_{-\sigma}^{\sigma} \widehat{f}(v - 2k\pi) \, \overline{\overline{g}}(v) dv \right| \leq n \|\widehat{g}\|_{L^\infty(\mathbb{R})} \int_{-\pi}^{\pi} \sum_{k \in \mathbb{Z}} \left| \widehat{f}(v - 2k\pi) \right| dv$$

$$= n \|\widehat{g}\|_{L^\infty(\mathbb{R})} \sum_{k \in \mathbb{Z}} \int_{(-2k-1)\pi}^{(-2k+1)\pi} \left| \widehat{f}(v) \right| dv = n \|\widehat{g}\|_{L^\infty(\mathbb{R})} \int_{-\infty}^{\infty} \left| \widehat{f}(v) \right| dv.$$

This inequality, with $\widehat{f}(\cdot)$ replaced by $\widehat{f}(\cdot - x) - \widehat{f}(\cdot - y)$, yields

$$\left| \sum_{k \in \mathbb{Z}} \int_{-\sigma}^{\sigma} \widehat{f}(v - x - 2k\pi) \, \overline{\overline{g}}(v) dv - \sum_{k \in \mathbb{Z}} \int_{-\sigma}^{\sigma} \widehat{f}(v - y - 2k\pi) \, \overline{\overline{g}}(v) dv \right|$$

$$\leq n \|\widehat{g}\|_{L^\infty(\mathbb{R})} \int_{-\infty}^{\infty} \left| \widehat{f}(v - x) - \widehat{f}(v - y) \right| dv,$$

and since the right side of this inequality tends to zero for $y \to x$ by the continuity in the mean of $\widehat{f} \in L^1(\mathbb{R})$, we have shown that the left-hand side of (3.9) is a continuous function of $x \in \mathbb{R}$. Thus, both sides being continuous, (3.9) must hold for all $x \in \mathbb{R}$.

$\square$

Proof of PDPS $\Rightarrow$ PSF for $f \in L^1(\mathbb{R})$ with $\widehat{f} \in S_1^1$. First assume $f \in F^1(\mathbb{R})$ with $\widehat{f} \in S_1^1$. We apply PDPS with $\sigma = \pi$, f replaced by $\widehat{f} \in F^1 \cap S_1^1 \subset F^2 \cap S_1^1$, and $g(t)$ replaced by $\widehat{g}_\rho(t)e^{-ixt} \in F^1$ of (2.12), (2.13). Noting that g_ρ, $\widehat{g}_\rho$ are real and even, (2.6), and

$$\overline{\left[\overline{\widehat{g}_\rho(t)e^{-ixt}}\right]^{\widehat{}}}(v) = \overline{\left[g_\rho(t-x)\right]^{\widehat{\widehat{}}}}(v) = \overline{g}_\rho(-v-x) = g_\rho(v+x),$$

$$\left[\overline{\widehat{g}_\rho(t)e^{-ixt}}\right]^{\widehat{}}(v) = \left[\widehat{g}_\rho(v)e^{ixt}\right]^{\widehat{}}(v) = \left[g_\rho(t+x)\right]^{\widehat{\widehat{}}}(v) = g_\rho(x-v),$$

we obtain for each $x \in \mathbb{R}$,

$$\sum_{k \in \mathbb{Z}} \widehat{f}(k)\widehat{g}_\rho(k)e^{ixk} = \sum_{k \in \mathbb{Z}} \int_{-\pi}^{\pi} \widehat{\widehat{f}}(v+2k\pi)g_\rho(v+x)\,dv$$

$$+ \sum_{k \in \mathbb{Z}} \widehat{f}(k)\frac{1}{\sqrt{2\pi}} \int_{|v| \geq \pi} g_\rho(x-v)e^{ikv}\,dv. \quad (3.11)$$

Now let $\rho \to 0+$. Since $\lim_{\rho \to 0+} \widehat{g}_\rho(k) = 1$ by (2.13), it follows that

$$\lim_{\rho \to 0+} \sum_{k \in \mathbb{Z}} \widehat{f}(k)\widehat{g}_\rho(k)e^{ixk} = \sum_{k \in \mathbb{Z}} \widehat{f}(k)e^{ixk} \quad (x \in \mathbb{R}),$$

where the interchange of summation and integration is allowed since the series is uniformly convergent with respect to $\rho > 0$ because of $\{\widehat{f}(k)\} \in \ell^1(\mathbb{Z})$ and $0 < \widehat{g}_\rho(k) \leq 1$ for all $k \in \mathbb{Z}$ and $\rho > 0$.

As to the right-hand side of (3.11), the first term can be rewritten as (cf. (2.6)),

$$\sum_{k \in \mathbb{Z}} \int_{-\pi}^{\pi} \widehat{\widehat{f}}(v+2k\pi)g_\rho(v+x)\,dv = \sum_{k \in \mathbb{Z}} \int_{-\pi}^{\pi} f(v-2k\pi)g_\rho(x-v)\,dv$$

$$= \int_{\mathbb{R}} \left\{ \sum_{k \in \mathbb{Z}} f(v+2k\pi) \right\} \operatorname{rect}(v)g_\rho(x-v)\,dv,$$

where the interchange of summation and integration is justified in view of the dominated convergence of the series. An application of Proposition 2 shows that the latter integral tends to $\sqrt{2\pi} \sum_{k \in \mathbb{Z}} f(v+2k\pi)$ a.e. on $(-\pi, \pi)$ for $\rho \to 0+$.

The integral in the second term on the right-hand side of (3.11) can be rewritten as

$$\frac{1}{\sqrt{2\pi}} \int_{\mathbb{R}} g_\rho(x-v)e^{ikv}\{1 - \operatorname{rect}(v)\}\,dv,$$

which tends to zero for $\rho \to 0+$ a.e. on $(-\pi, \pi)$ by Proposition 2. Hence we obtain

$$\lim_{\rho \to 0+} \sum_{k \in \mathbb{Z}} \widehat{f}(k)\frac{1}{\sqrt{2\pi}} \int_{|v| \geq \pi} g_\rho(v-x)e^{ikv}\,dv = 0 \quad (\text{a.e. on } (-\pi, \pi)),$$

noting again that $\{\widehat{f}(k)\} \in \ell^1(\mathbb{Z})$.

Altogether it follows from (3.11) for $\rho \to 0+$,

$$\sum_{k \in \mathbb{Z}} \widehat{f}(k) e^{ixk} = \sqrt{2\pi} \sum_{k \in \mathbb{Z}} f(v + 2k\pi) \quad \left(\text{a.e. on } (-\pi, \pi)\right). \qquad (3.12)$$

Since both sides of (3.12) define a 2π-periodic function, equation (3.12) holds even a. e. on $\mathbb{R}$, which is PSF for the space $\{f \in F^1; \widehat{f} \in S_1^1\} \subset L^1(\mathbb{R})$.

In order to extend PSF to the whole of $L^1(\mathbb{R})$ with $\widehat{f} \in S_1^1$, we have only to note the remarks on PSF in Sec. 2.4, since $\{f \in F^1; \widehat{f} \in S_1^1\}$ is dense in $L^1(\mathbb{R})$. For example, the convolution $f_\rho := f * g_\rho$ of $f \in L^1(\mathbb{R})$ with g_ρ of (2.12) belongs to F^1, $\widehat{f_\rho} \in S_1^1$ for all $\rho > 0$, and $\lim_{\rho \to \infty} f_\rho = f$ in $L^1(\mathbb{R})$-norm (Proposition 2).   □

The equivalence PSF $\Leftrightarrow$ GPDF was established in [18] with the PSF-version

$$\sqrt{2\pi} \sum_{k \in \mathbb{Z}} \widehat{f}(x + 2k\pi w) = \sum_{k \in \mathbb{Z}} f\left(\frac{k}{w}\right) e^{ikx/w} \quad (x \in \mathbb{R}).$$

whereby the roles of f and $\widehat{f}$ have been exchanged, under the additional assumption $\widehat{f} \in BV(\mathbb{R})$.

The present proof is modeled on that of [18], but the lack of the BV-assumption in our case causes some additional arguments.

3.4 PSF and ASF

As to the equivalence of PSF and ASF, it was already established in [23, 24], however with different versions of PSF.

Proof of PSF $\Rightarrow$ ASF for $f \in F^1 \cap S_1^1$. We again apply PSF (1.10) to the function $\widehat{f} \in F^1$, noting that $\left(\widehat{\widehat{f}}(k)\right) = \left(f(-k)\right) \in \ell^1(\mathbb{Z})$, giving

$$\sqrt{2\pi} \sum_{k \in \mathbb{Z}} \widehat{f}(v + 2k\pi) = \sum_{k \in \mathbb{Z}} f(k) e^{-ikv} \quad a.e.,$$

Hence, recalling (2.7),

$$\frac{1}{\sqrt{2\pi}} \int_{-\pi}^{\pi} \sum_{k \in \mathbb{Z}} \widehat{f}(v + 2k\pi) \, dv = \frac{1}{2\pi} \int_{-\pi}^{\pi} \sum_{k \in \mathbb{Z}} f(k) e^{-ikv} e^{ivt} \, dv$$

$$= \frac{1}{2\pi} \sum_{k \in \mathbb{Z}} f(k) \int_{-\pi}^{\pi} e^{-ikv} e^{ivt} \, dv = \frac{1}{2\pi} \sum_{k \in \mathbb{Z}} f(k) \int_{-\infty}^{\infty} \text{rect}(v) e^{-ikv} e^{ivt} \, dv$$

$$= \sum_{k \in \mathbb{Z}} f(k) \, \text{sinc}(t - k) = (S_1 f)(t) \quad (t \in \mathbb{R}),$$

the interchange of summation and integration being justified in view of the absolute and uniform convergence of the series involved.

On the other hand, we have

$$\int_{-\pi}^{\pi} \sum_{k \in \mathbb{Z}} \widehat{f}(v + 2k\pi) e^{ivt} \, dv = \sum_{k \in \mathbb{Z}} \int_{(2k-1)\pi}^{(2k+1)\pi} \widehat{f}(v) e^{it(v-2k\pi)} \, dv \qquad (t \in \mathbb{R}),$$

yielding

$$(S_1 f)(t) = \frac{1}{\sqrt{2\pi}} \sum_{k \in \mathbb{Z}} e^{-i2k\pi t} \int_{(2k-1)\pi}^{(2k+1)\pi} \widehat{f}(v) e^{ivt} \, dv \qquad (t \in \mathbb{R}). \tag{3.13}$$

But by the Fourier inversion integral (2.6),

$$f(t) = \frac{1}{\sqrt{2\pi}} \int_{\mathbb{R}} \widehat{f}(v) e^{ivt} \, dv = \frac{1}{\sqrt{2\pi}} \sum_{k \in \mathbb{Z}} \int_{(2k-1)\pi}^{(2k+1)\pi} \widehat{f}(v) e^{ivt} \, dv \quad (t \in \mathbb{R}). \tag{3.14}$$

Subtracting (3.13) from (3.14) yields $f(t) - (S_1 f)(t) = (R_1 f)(t)$, thus the approximate sampling theorem. $\qquad\square$

Proof of ASF $\Rightarrow$ PSF Noting (3.14) it follows immediately from ASF with remainder (1.2) that for $f \in F^2$ and $w = 1$

$$\sum_{k \in \mathbb{Z}} f(k) \operatorname{sinc}(t - k) = \frac{1}{\sqrt{2\pi}} \sum_{k \in \mathbb{Z}} e^{-i2k\pi t} \int_{(2k-1)\pi}^{(2k+1)\pi} \widehat{f}(v) e^{ivt} \, dv \quad (t \in \mathbb{R}). \tag{3.15}$$

Using (2.5) and (2.7) one can rewrite the terms on the right-hand side as a convolution product, namely

$$\frac{e^{-i2k\pi t}}{\sqrt{2\pi}} \int_{(2k-1)\pi}^{(2k+1)\pi} \widehat{f}(v) e^{itv} \, dv = \frac{e^{-i2k\pi t}}{\sqrt{2\pi}} \int_{\mathbb{R}} \widehat{f}(v) \operatorname{rect}(v - 2k\pi) e^{itv} \, dv$$

$$= \int_{\mathbb{R}} f(u) e^{-i2k\pi u} \operatorname{sinc}(t - u) \, du = \sqrt{2\pi} \big[(f(\cdot) e^{-i2k\pi \cdot}) * \operatorname{sinc}(\cdot) \big](t) \quad (t \in \mathbb{R}).$$

For the following, we first assume $f \in \widehat{B}_\sigma^1$ for some $\sigma > 0$. Then the terms on the right of (3.15) vanish for $|k| > \sigma_0 := (\sigma/\pi + 1)/2$, and (3.15) can be rewritten as

$$\sum_{k \in \mathbb{Z}} f(k) \operatorname{sinc}(t - k) = \sqrt{2\pi} \sum_{|k| \le \sigma_0} \big[(f(\cdot) e^{-i2k\pi \cdot}) * \operatorname{sinc}(\cdot) \big](t) \quad (t \in \mathbb{R}).$$

Now, taking the L^2-Fourier transform of this equation, noting that the series on the left converges with respect to $L^2(\mathbb{R})$-norm (Lemma 3), Proposition 1 and (2.8), we obtain

$$\frac{1}{\sqrt{2\pi}} \sum_{k \in \mathbb{Z}} f(k) \operatorname{rect}(v) e^{ikv} = \sum_{|k| \le \sigma_0} \widehat{f}(v + 2k\pi) \operatorname{rect}(v) \quad a.e. \tag{3.16}$$

Since $f \in \widehat{B}_\sigma^1$, $\hat{f}$ can be regarded as the L^1-Fourier transform, which is continuous. Then (3.16) holds even for every v in a neighbourhood of the origin, since all functions involved are continuous there. So one may take $v = 0$ to obtain

$$\frac{1}{\sqrt{2\pi}} \sum_{k \in \mathbb{Z}} f(k) = \sum_{|k| \leq \sigma_0} \hat{f}(2k\pi).$$

Replacing now $f(\cdot)$ by $2\pi f(x + 2\pi \cdot)$, and noting Proposition 1 (a) gives PSF.

It remains to remove the restriction $f \in \widehat{B}_\sigma^1$ for some $\sigma > 0$. According to Sect. 2.4 it suffices to show that $\bigcup_{\sigma > 0} \widehat{B}_\sigma^1$ is dense in $L^1(\mathbb{R})$. To see this, one may take $f * \chi_\rho, \rho > 0$, with $f \in L^1(\mathbb{R})$ and χ_ρ of (2.14). This convolution belongs to $\widehat{B}_{2\pi\rho}^1$, noting (2.15) and Proposition 1 (b), and $\lim_{\rho \to \infty} f * \chi_\rho = f$ in $L^1(\mathbb{R})$-norm (Proposition 2 (b). □

4 PSF and the functional equation for $\zeta(s)$

First some remarks concerning the ζ-function. Since the series (1.15) is locally uniformly convergent in the half-plane $\sigma > 1$, $\zeta(s)$ is analytic there. To extend it to a function which is analytic in the whole complex plane $\mathbb{C}$ except $s = 1$, one can employ the representation

$$\zeta(s) = s \int_1^\infty u^{-s-1} \left(\mu(u) + \frac{1}{2} \right) du + \frac{1}{s-1} + \frac{1}{2} \qquad (4.1)$$

with the measure $\mu(u) := \lfloor u \rfloor - u$, where $\lfloor u \rfloor$ denotes the largest integer less than or equal to u; see Titchmarsh [82, (2.1.4)].

Noting that $| \int_1^u (\mu(t) + \frac{1}{2}) dt |$ is bounded by $\frac{1}{8}$ on $\mathbb{R}$, one can easily show by partial integration that the integral is locally uniformly convergent in the half-plane $\sigma > -1$. Hence (4.1) defines an extension of (1.15) to a function which is analytic for $\sigma > -1$ except for the point $s = 1$, where there is a simple pole with residue 1. This function can be continued further to a function which is analytic in the whole complex plane except for $s = 1$.

From (4.1) one can also deduce by partial integration the estimate

$$|\zeta(s)| \leq |s| + \frac{1}{|s-1|} + \frac{1}{2} \quad (s = \sigma + i\tau \in \mathbb{C} \setminus \{1\}, \sigma > -1). \qquad (4.2)$$

In the strip $0 < \sigma < 1$ the representation (4.1) can be rewritten as (see Tichmarsh (2.1.5))

$$\zeta(s) = s \int_0^\infty u^{-s-1} \mu(u) du \quad (s = \sigma + i\tau, 0 < \tau < 1), \qquad (4.3)$$

noting that

$$\int_0^1 u^{-s-1} \mu(u) du = \int_0^1 u^{-s} du = \frac{1}{s-1}.$$

Now we turn to the equivalence PSF $\Leftrightarrow$ FERZ.

Proof of PSF $\Rightarrow$ FERZ If f belongs to the Schwartz space $\mathcal{S}$ of rapidly decreasing functions, the PSF (1.10), in its simplest form for $w = 1$, $x = 0$, reads

$$\sqrt{2\pi} \sum_{k \in \mathbb{Z}} f(2k\pi) = \sum_{k \in \mathbb{Z}} \widehat{f}(k).$$

In fact, under the given assumption $f \in \mathcal{S}$, Eq. (1.10) holds for all $x \in \mathbb{R}$, since both series are uniformly convergent, at least on compact subsets of $\mathbb{R}$, and hence define continuous functions.

If f is even, so will be $\widehat{f}$, and PSF turns into

$$\sqrt{2\pi} \left\{ \sum_{k=1}^{\infty} f(2\pi k) - \int_0^{\infty} f(2\pi u)\, du \right\} = \int_0^{\infty} \widehat{f}(v)\, dv - \sum_{k=1}^{\infty} \widehat{f}(k),$$

noting that $\frac{1}{2}\widehat{f}(0) = (1/\sqrt{2\pi}) \int_0^{\infty} f(x)\, dx = \sqrt{2\pi} \int_0^{\infty} f(2\pi u)\, du$ and, by the Fourier inversion formula (2.6), $\frac{1}{2}\sqrt{2\pi} f(0) = \int_0^{\infty} \widehat{f}(v)\, dv$.

This equation can be conveniently be written as

$$\sqrt{2\pi} \int_0^{\infty} f(2\pi u)\, d\mu(u) = \int_0^{\infty} \widehat{f}(v)\, d\mu(v), \tag{4.4}$$

Further, as $f \in \mathcal{S}$ implies f', $\widehat{f}' \in L^1(\mathbb{R})$ and $\lim_{u \to \infty} f(u) = \lim_{v \to \infty} \widehat{f}(v) = 0$, the two integrals can be converted into ordinary Lebesgue integrals by partial integration, namely,

$$\sqrt{2\pi} \int_0^{\infty} f(2\pi u)\, d\mu(u) = \sqrt{2\pi}\, f(2\pi u)\mu(u) \Big|_0^{\infty} - (2\pi)^{3/2} \int_0^{\infty} \mu(u) f'(2\pi u)\, du$$

$$= -(2\pi)^{3/2} \int_0^{\infty} \mu(u) f'(2\pi u)\, du,$$

and similarly,

$$\int_0^{\infty} \widehat{f}(v)\, d\mu(v) = -\int_0^{\infty} \mu(v) \widehat{f}'(v)\, dv.$$

Hence we obtain the following "integrated form" of PSF,

$$(2\pi)^{3/2} \int_0^{\infty} \mu(u) f'(2\pi u)\, du = \int_0^{\infty} \mu(v) \widehat{f}'(v)\, dv. \tag{4.5}$$

For the Gauß–Weierstraß kernel g_ρ of (2.12), (2.13) with $\rho = \pi t^2$, $t > 0$, this identity reads,

$$-2\pi t^{-3} \int_0^{\infty} u\, \mu(u) e^{-\pi t^{-2} u^2}\, du = -2\pi t^2 \int_0^{\infty} v\, \mu(v) e^{-\pi t^2 v^2}\, dv. \tag{4.6}$$

Now we multiply both sides of (4.6) by $-t^{-\sigma}$, $\sigma > 0$, integrate with respect to t from 0 to ∞, and interchange the order of integration, which is allowed by the Fubini–Tonelli theorem [40, (21.13)]. This yields for the left-hand side by a simple change of variables

$$2\pi \int_0^\infty u\,\mu(u) \int_0^\infty t^{-3-\sigma} e^{-\pi t^{-2}u^2}\,dt\,du = \frac{1}{2}\pi^{-\sigma/2}\Gamma\left(\frac{\sigma}{2}\right)\sigma \int_0^\infty u^{-\sigma-1}\mu(u)\,du,$$

and for the right-hand side

$$2\pi \int_0^\infty v\,\mu(v) \int_0^\infty t^{2-\sigma} e^{-\pi t^2 v^2}\,dt\,dv$$
$$= \frac{1}{2}\pi^{-\frac{1-\sigma}{2}}\Gamma\left(\frac{1-\sigma}{2}\right)(\sigma-1)\int_0^\infty v^{\sigma-2}\mu(v)\,dv.$$

Using the representation (4.3) for the ζ-function, the functional equation follows for $s = \sigma$ with $0 < \sigma < 1$, and analytic continuation yields the general result. □

Proof of FERZ $\Rightarrow$ PSF It is well-known that an infinite series can be transformed into an integral involving $\zeta(s)$ (as well as vice versa!) using Mellin transform methods; see e. g. [19]. Formally one proceeds as follows: To sum

$$g(x) := \sum_{k=1}^\infty f(kx) \quad (x \in \mathbb{R}_+),$$

one first applies the Mellin transform

$$\mathcal{M}(f)(s) \equiv \widehat{f_M}(s) := \int_0^\infty f(u)u^{s-1}\,du \quad (s = \sigma + i\tau \in \mathbb{C})$$

on both sides to yield

$$\widehat{g_M}(s) = \sum_{k=1}^\infty \frac{1}{k^s}\widehat{f_M}(s) = \widehat{f_M}(s)\zeta(s), \tag{4.7}$$

since the Mellin transform of $f(kx)$, where $k > 0$, is $k^{-s}\widehat{f_M}(s)$. The Mellin inversion formula

$$f(x) = \frac{1}{2\pi i}\int_{c-i\infty}^{c+i\infty} \widehat{f_M}(s)x^{-s}\,ds \quad (c > 0)$$

would then lead to

$$g(x) = \sum_{k=1}^\infty f(kx) = \frac{1}{2\pi i}\int_{c-i\infty}^{c+i\infty} \widehat{f_M}(s)\zeta(s)x^{-s}\,ds \quad (c > 1). \tag{4.8}$$

In our situation, to work with the sum (4.8) we need to move the integration contour from $c + it$ to $1 - c + it$, $-\infty < t < +\infty$. The equation for $x = 2\pi$ then becomes

$$\sum_{k=1}^{\infty} f(2k\pi) = \frac{1}{2\pi i} \int_{1-c-i\infty}^{1-c+i\infty} \widehat{f}_M(s)\zeta(s)(2\pi)^{-s}\,ds + S,$$

where S accounts for the sum of the residues of the integrand due to the contour shift. The change of variable $s \mapsto 1 - s$ in the integral causes the appearance of a term $\zeta(1 - s)$ in the integrand. The functional equation is used at this point, to make the term $\zeta(s)$ reappear. But, as Eq. (4.8) shows, the inverse Mellin transform of a product involving $\zeta(s)$ represents a series $\sum h(k)$, for some function h. This function will turn out to be the Fourier transform of f, so that

$$\sum_{k=1}^{\infty} f(2k\pi) = \sum_{k=1}^{\infty} f^{\wedge}(k) + S.$$

A computation will show that the residues are $R_0 = -\frac{1}{2}f(0)$ and $R_1 = \frac{1}{2\pi}\int_0^{\infty} f(x)\,dx$, and PSF will then follow.

To carry out the above sketch of the proof and to justify these steps, consider a locally integrable function $f(x)$. Let it be $\mathcal{O}(x^{-a})$ as $x \to 0$ and $\mathcal{O}(x^{-b})$ as $x \to \infty$ for some $a < b$. The Mellin transform $\widehat{f}_M(s)$ then exists for any complex $s = \sigma + i\tau$ in the fundamental strip $a < \sigma < b$. This follows from

$$\left| \int_0^{\infty} f(u)u^{s-1}du \right| \leq \int_0^1 |f(u)|u^{\sigma-1}du + \int_1^{\infty} |f(u)|u^{\sigma-1}du$$

$$\leq A \int_0^1 u^{-a+\sigma-1}du + B \int_1^{\infty} u^{-b+\sigma-1}du,$$

since the first integral exists if $\sigma > a$ and the second if $\sigma < b$. If, in addition, $f(y)$ is of bounded variation in a neighbourhood of $y = x$, the Mellin inversion formula holds, in the form

$$\frac{f(x^+) + f(x^-)}{2} = \frac{1}{2\pi i} \int_{\sigma-i\infty}^{\sigma+i\infty} \widehat{f}_M(s)x^{-s}\,ds,$$

for $a < \sigma < b$. To justify it one may appeal to the relation between the Mellin and the Fourier transform, which is revealed by the change of variables $x = e^{-y}$, since, as we have seen, $f(x)x^{\sigma-1} \in L^1(0, \infty)$ under the stated hypothesis.

Now, if the intersection of the fundamental strip $a < \sigma < b$ with the half-plane $\sigma > 1$ is nonempty, and if s belongs to that intersection, then (4.7) holds because there exists under the stated conditions

$$\sum_{k=1}^{\infty} \left| \frac{1}{k^s} \right| \int_0^{\infty} |u^{s-1}f(u)|du.$$

If g is a continuous function of bounded variation in a neighbourhood of x, the application of the Mellin inversion formula to (4.7) then leads to the identity (4.8), that is,

$$g(x) = \sum_{k=1}^{\infty} f(kx) = \frac{1}{2\pi i} \int_{\sigma-i\infty}^{\sigma+i\infty} \widehat{f}_M(s)\zeta(s)x^{-s}\,ds.$$

To continue with the proof, let f be an even function in the Schwartz space $\mathcal{S}$. Then, the Mellin transform of f exists for all $s = \sigma + it$ with $\sigma > 0$. Below we will have to integrate $\widehat{f}_M(s)$ in a region contained in the left semiplane, so we need to investigate its analytic extension there. The necessary information can be obtained through an integration by parts of the Mellin integral,

$$\widehat{f}_M(s) = -\frac{1}{s} \int_0^{\infty} f'(x)x^s\,dx = -\frac{1}{s}\widehat{f'}_M(s+1).$$

The procedure can be repeated to give

$$\widehat{f}_M(s) = \frac{(-1)^r}{s(s+1)\cdots(s+r-1)}\widehat{[f^{(r)}]}_M(s+r) \quad (r \in \mathbb{N}). \tag{4.9}$$

This provides the desired analytic extension and shows that $\widehat{f}_M(s)$ may have simple poles at the non-positive integers. The residue at $s = 0$ is given by

$$\lim_{s\to 0} s\left(-\frac{1}{s}\widehat{f'}_M(s+1)\right) = -\widehat{f'}_M(1) = -\int_0^{\infty} f'(x)\,dx = f(0). \tag{4.10}$$

Although we will not need any other residues, one could easily check that the residue at $s = -n$ is $\frac{1}{n!}f^{(n)}(0)$.

We are now ready to return to the fundamental transformation (4.8) with $x = 2\pi$. We take an arbitrary $1 < c < 2$ and change the contour,

$$\frac{1}{2\pi i} \int_{c-i\infty}^{c+i\infty} \widehat{f}_M(s)\zeta(s)(2\pi)^{-s}\,ds = \frac{1}{2\pi i} \int_{1-c-i\infty}^{1-c+i\infty} \widehat{f}_M(s)\zeta(s)(2\pi)^{-s}\,ds + S,$$

where S is the sum of the residues of $\widehat{f}_M(s)\zeta(s)(2\pi)^{-s}$ at the poles in the strip $1 - c < \sigma < c$. There are two such poles: one at $s = 1$, due to $\zeta(s)$, and another at $s = 0$, due to $\widehat{f}_M(s)$. The residue of the integrand at the former is

$$R_1 = \frac{1}{2\pi}\widehat{f}_M(1) = \frac{1}{2\pi} \int_0^{\infty} f(x)\,dx$$

and at the latter

$$R_0 = f(0)\zeta(0) = -\frac{1}{2}f(0),$$

in view of (4.10). As a result, (4.8) becomes

$$\sum_{k=1}^{\infty} f(2k\pi) = \frac{1}{2\pi i} \int_{1-c-i\infty}^{1-c+i\infty} \widehat{f_M}(s)\zeta(s)(2\pi)^{-s}\,ds + \frac{1}{2\pi}\int_0^{\infty} f(x)\,dx - \frac{1}{2}f(0).$$

We now replace $s \mapsto 1 - s$ in the integral

$$\sum_{k=1}^{\infty} f(2k\pi)$$

$$= \frac{1}{2\pi}\int_0^{\infty} f(x)\,dx - \frac{1}{2}f(0) + \frac{1}{2\pi i}\int_{c-i\infty}^{c+i\infty} \widehat{f_M}(1-s)\zeta(1-s)(2\pi)^{s-1}\,ds$$

and use the functional equation in the form (1.16) to deduce

$$\sum_{k=1}^{\infty} f(2k\pi) \tag{4.11}$$

$$= \frac{1}{2\pi}\int_0^{\infty} f(x)\,dx - \frac{1}{2}f(0) + \frac{1}{2\pi i}\int_{c-i\infty}^{c+i\infty} \widehat{f_M}(1-s)2\cos\left(\frac{\pi s}{2}\right)\Gamma(s)\zeta(s)\,ds. \tag{4.12}$$

According to the principle expressed by (4.8), the last integral represents the sum of a series,

$$\frac{1}{2\pi i}\int_{c-i\infty}^{c+i\infty} \widehat{f_M}(1-s)2\cos\left(\frac{\pi s}{2}\right)\Gamma(s)\zeta(s)\,ds = \sum_{k=1}^{\infty} h(k). \tag{4.13}$$

The question is: what is the relation between h and f? The elementary Mellin transform formula

$$\int_0^{\infty} \cos(ax)x^{s-1}\,dx = a^{-s}\cos\left(\frac{\pi s}{2}\right)\Gamma(s) \qquad (0 < \Re s < 1) \tag{4.14}$$

shows that

$$\int_0^{\infty} \widehat{f}(v)v^{s-1}\,dv = 2\int_0^{\infty}\left(\frac{1}{\sqrt{2\pi}}\int_0^{\infty} f(t)\cos(vt)\,dt\right)v^{s-1}\,dv$$

$$= \frac{2}{\sqrt{2\pi}}\int_0^{\infty} f(t)\left(\int_0^{\infty}\cos(vt)v^{s-1}\,dv\right)dt$$

$$= \frac{2}{\sqrt{2\pi}}\cos\left(\frac{\pi s}{2}\right)\Gamma(s)\int_0^{\infty} f(t)t^{-s}\,dt$$

$$= \frac{2}{\sqrt{2\pi}}\cos\left(\frac{\pi s}{2}\right)\Gamma(s)\widehat{f_M}(1-s). \tag{4.15}$$

In order to justify the interchange of the integration order, first note that

$$I := \int_0^\infty f(t) \left(\int_0^\infty \cos(vt) v^{s-1} \, dv \right) dt$$
$$= \int_0^\infty f(t) \lim_{R \to \infty} \left(\int_0^R \cos(vt) v^{s-1} \, dv \right) dt$$
$$= \lim_{R \to \infty} \int_0^\infty f(t) \left(\int_0^R \cos(vt) v^{s-1} \, dv \right) dt. \qquad (4.16)$$

Here the interchange of the limit with the outer integral is justified by Lebesgue's dominated convergence theorem, since

$$\left| f(t) \int_0^R \cos(vt) v^{s-1} \, dv \right| = \left| t^{-s} f(t) \int_0^{Rt} \cos(u) u^{s-1} \, du \right|$$
$$\leq c(s) t^{-\sigma} |f(t)| \in L^1(0, \infty),$$

where the constant $c(s)$ is an upper bound for the continuous and bounded function $x \mapsto \left| \int_0^x \cos(u) u^{s-1} \, du \right|$ on $[0, \infty)$.

Furthermore, the order of integration in the last line of (4.16) can be interchanged by Fubini's theorem, and one obtains

$$I = \lim_{R \to \infty} \int_0^R \left(\int_0^\infty f(t) \cos(vt) dt \right) v^{s-1} \, dv,$$

which proves the desired interchange of integrals in (4.15).

Hence we have shown that

$$\int_0^\infty \widehat{f}(v) v^{s-1} \, dv = \frac{2}{\sqrt{2\pi}} \cos\left(\frac{\pi s}{2}\right) \Gamma(s) \widehat{f}_M(1 - s). \qquad (4.17)$$

Since we have used the elementary formula (4.14) to deduce this equation, it holds at first for $0 < \Re s < 1$. On the other hand, both sides of (4.17) define analytic functions for $\Re z > 0$, and hence (4.17) is even valid for $\Re z > 0$.

Now, equation (4.17) reveals that the factor that multiplies $\zeta(s)$ in the integrand of (4.13) is the Mellin transform of the Fourier transform $\widehat{f}$ of f. Thus (4.13) becomes

$$\frac{1}{2\pi i} \int_{c-i\infty}^{c+i\infty} \widehat{f}_M(1 - s) 2 \cos\left(\frac{\pi s}{2}\right) \Gamma(s) \zeta(s) \, ds = \frac{1}{\sqrt{2\pi}} \sum_{k=1}^\infty \widehat{f}(k).$$

Replacing it in (4.11) finally yields

$$\sum_{k=1}^\infty f(2k\pi) = \frac{1}{2\pi} \int_0^\infty f(x) \, dx - \frac{1}{2} f(0) + \frac{1}{\sqrt{2\pi}} \sum_{k=1}^\infty \widehat{f}(k).$$

Noting that $\frac{1}{2\pi}\int_0^\infty f(x)\,dx = \frac{1}{2}\frac{1}{\sqrt{2\pi}}\,\widehat{f}(0)$, the PSF (1.10) for even functions in the Schwarz space $\mathcal{S}$ and $x = 0$ follows.

To extend this to arbitrary functions $f \in \mathcal{S}$, one decomposes f in the form $f(t) = \frac{1}{2}[f(t) + f(-t)] + \frac{1}{2}[f(t) - f(-t)]$, i.e., in an even and an odd part. Since both sums in (1.10) vanish for odd functions and $x = 0$, PSF for $f \in \mathcal{S}$ and $x = 0$ follows. To obtain PSF for arbitrary $x \in \mathbb{R}$, one replaces $f(\cdot)$ by $2\pi f(x + 2\pi \cdot)$ as in the proof of ASF $\Rightarrow$ PSF. Finally, one uses the fact that $\mathcal{S}$ is dense in $L^1(\mathbb{R})$ to deduce the general PSF for $f \in L^1(\mathbb{R})$; cf. Sect. 2.4.   □

Let us remark that a proof of PSF yields FERZ is to be found in Mordell [57], and the converse in Ferrar [30]. In his treatise Titchmarsh [82] presented seven different methods of proof of the functional equation; see also Kahane and Mandelbrojt [50], Bellman [8], Ivić [49], Patterson [66, Chapter 2], Karatsuba and Voronin [51], Rooney [75], Flajolet et al. [31], Brüdern [13, pp. 58–66], Newman [63]. Latter proofs are Knopp and Robins [55], Schuster [78], Murty [58] and Higgins [45].

Whereas several of the above mentioned books or papers contain a proof of PSF $\Rightarrow$ FERZ, proofs of the converse FERZ $\Rightarrow$ PSF, an important part of our procedure, are rarer. In fact, there are other proofs of FERZ $\Leftrightarrow$ PSF, but ours are fully detailed and more than just polished versions of others. Especially the delicate interchanges of the orders of summation and/or integration needed in the proofs are justified in great detail.

In his exhaustive report [44] on four papers of Hamburger [33–36] of 1921–1922 in connection with Riemann's functional equation, R. Higgins gave a complete review of Hamburger's collection of five equivalent results, namely that Riemann's functional equation for the zeta function, Jacobi's transformation formula for the elliptic theta function, the partial fractions expansion for the cotangent function, Poisson's summation formula and a special Fourier series are all equivalent. He also discussed these in the light of later contributions.

Here the basic earlier work being the "Lehrbuch der Thetafunktionen" by Krazer [56], papers by Siegel [79], those of Mordell [57] and Ferrar [30], as well as more results in the light of work by Doetsch [27] and Klusch [52,53]. Klusch proves many implications, and collects others, between well-known results in signal analysis, number theory and applied mathematics. In particular, he gives a direct proof of FERZ implies ASF, an open question at the time. These results do not seem to be at all well known.

5 A proof of ASF

In the previous section we have shown that the formulae of Sects. 1 or 3 can all be deduced from each other. Hence it suffices to prove one of them in order to verify all. We will now prove ASF under the weaker hypothesis that $f \in F^2$ and with a weaker conclusion on the convergence of the series. From that result, the statement (1.6) for $f \in F^2 \cap S_w^1$ given in Sect. 1 is easily deduced.

Theorem 1 *For $w > 0$ and $f \in F^2$, we have*

$$f(t) = \sum_{k \in \mathbb{Z}} f\left(\frac{k}{w}\right) \operatorname{sinc}(wt - k) + R_w f(t),$$

where the series converges pointwise as the symmetric sum.

Proof It suffices to consider the case $w = 1$. Employing the Fourier inversion theorem for $f \in F^2$, see [20, p. 214, Proposition 5.2.16], we may rewrite the remainder (1.2) as

$$R_1 f(t) = \frac{1}{\sqrt{2\pi}} \left[\int_{\mathbb{R}} \widehat{f}(v) e^{ivt} \, dt - \sum_{k \in \mathbb{Z}} \int_{(2k-1)\pi}^{(2k+1)\pi} \widehat{f}(v) e^{it(v - 2\pi k)} \, dv \right]$$

$$= f(t) - \frac{1}{\sqrt{2\pi}} \int_{\mathbb{R}} \widehat{f}(v) g(t, v) \, dv, \tag{5.1}$$

where $g(t, \cdot)$ is the function obtained by restricting $e^{it \cdot}$ to the interval $[-\pi, \pi)$ and extending it to $\mathbb{R}$ by 2π periodic continuation. By a simple calculation we obtain the Fourier expansion

$$g(t, v) = \sum_{k \in \mathbb{Z}} \operatorname{sinc}(t - k) e^{ikv}. \tag{5.2}$$

Since $g(t, \cdot)$ is of bounded variation, the Fourier series converges and (5.2) holds at each point of continuity. Moreover, the partial sums

$$g_N(t, v) := \sum_{|k| \leq N} \operatorname{sinc}(t - k) e^{ikv}$$

are uniformly bounded, that is

$$|g_N(t, v)| \leq C \qquad (v \in \mathbb{R}, N \in \mathbb{N})$$

for some constant C; see [85, p. 90, Thm. 3.7]. Since $\widehat{f} \in L^1(\mathbb{R})$, we see that $\widehat{f}(\cdot) g_N(t, \cdot)$ has an absolutely integrable majorant for all $N \in \mathbb{N}$. Therefore, Lebesgue's dominated convergence theorem allows us to conclude that

$$\frac{1}{\sqrt{2\pi}} \int_{\mathbb{R}} \widehat{f}(v) g(t, v) \, dv = \lim_{N \to \infty} \frac{1}{\sqrt{2\pi}} \int_{\mathbb{R}} \widehat{f}(v) g_N(t, v) \div$$

$$= \lim_{N \to \infty} \sum_{|k| \leq N} \left\{ \frac{1}{\sqrt{2\pi}} \int_{\mathbb{R}} \widehat{f}(v) e^{ikv} \, dv \right\} \operatorname{sinc}(t - k)$$

$$= \lim_{N \to \infty} \sum_{|k| \leq N} f(k) \operatorname{sinc}(t - k),$$

where again the Fourier inversion theorem (2.6) has been used in the last step. Now the proof is completed by substituting the last expression in (5.1).   □

Let us mention that in the meantime some of the authors have shown that ASF is equivalent to the classical sampling theorem (1.7); see [14]. The above proof of ASF is, on the other hand, a Fourier analytic proof, fully independent of CSF and all the theorems of this paper.

6 Epilogue: some possible generalizations

The natural question arises as to whether the results above can be extended to higher dimensions or other settings. There are (at least) two ways : via functional analysis, based on r.k. theory, and via abstract harmonic analysis, using locally compact abelian (LCA) groups.

In the first of these generalizations, sampling itself is placed in the more general context of reproducing kernel Hilbert spaces of functions defined on an abstract set. In the second, the real line $\mathbb{R}$, which can be thought of as time domain, is replaced by an abstract LCA topological group G, as usual assumed to be Hausdorff. Functions are now defined on a LCA group, itself a fruitful generalization of the real line.

6.1 Reproducing kernel theory

The purpose of this section is to give some very brief and necessarily quite simplified background to Hilbert spaces with reproducing kernel. The material in this section, proofs etc., can be found in [42,77]. There are several studies in the literature of sampling to be found in the general area of reproducing kernel spaces, and we can ask whether results of the present paper might be given more general forms in that theory. For example, Nashed and his associates have given sampling theorems in certain Hilbert, Banach, Sobolev and translation invariant spaces, and addressed such probems as the construction of reproducing kernel Hilbert and Banach spaces with a given sampling set (see, e. g., [59]). Much of this work is quite technical and uses, e. g., iterative methods. Saitoh's theory of linear transformations of Hilbert space offers another possibility for this kind of extension (see, e. g., [77]).

6.1.1 Hilbert spaces with reproducing kernel

Let K be a separable Hilbert space with inner product denoted by $\langle \cdot, \cdot \rangle_K$ and norm denoted by $\| \cdot \|_K$.

Definition 1 Let K consist of complex valued functions defined on E, an abstract set. Then K is said to have *reproducing kernel* if there exists a function $k : E \times E \mapsto \mathbb{C}$, such that $k(\cdot, t) \in K$ for every $t \in E$, and the *reproducing equation*

$$f(t) = \langle f, k(\cdot, t) \rangle_K$$

holds for every $f \in \mathrm{K}$. Such a space is called a *reproducing kernel Hilbert space (RKHS)* and $k(s, t)$ is its *reproducing kernel*.

6.1.2 A summary of the relevant Saitoh theory

We now quote that part of Saitoh's fundamental theory of linear transformations of Hilbert space which provides a suitable setting for the present discussion ([77, Ch. 2]), but first some background is needed.

Let $\mathfrak{H}$ be a separable Hilbert space with inner product denoted by $\langle \cdot, \cdot \rangle_{\mathfrak{H}}$, and for each t belonging to an abstract set E, let κ_t be a mapping of E into $\mathfrak{H}$. Then $k(s, t) := \langle \kappa_t, \kappa_s \rangle_{\mathfrak{H}}$ is defined on $E \times E$ and is called the *kernel function* of the map κ_t. This kernel function is a *positive matrix* [77, Ch. 2, Sect. 2] and as such it determines one and only one Hilbert space for which it is the reproducing kernel. This Hilbert space is denoted by R_k; it turns out to be the set of images of $\mathfrak{H}$ under the transformation

$$(\mathcal{L}\varphi)(t) := \langle \varphi, \kappa_t \rangle_{\mathfrak{H}}, \qquad (\varphi \in \mathfrak{H}) \tag{6.1}$$

and has reproducing kernel $k(s, t) = (\mathcal{L}\kappa_t)(s)$. This situation is governed by:

Theorem 2 (Saitoh) *With the notations established above, R_k is a Hilbert space which has the reproducing kernel $k(\cdot, \cdot)$, and is uniquely determined by this kernel. We have, for $f \in R_k$,*

$$\|f\|_{R_k} = \|\mathcal{L}\omega\|_{R_k} \leq \|\omega\|_{\mathfrak{H}}, \tag{6.2}$$

and there exists a unique member, ω_0 say, of the class of all ω's satisfying (6.2) such that

$$f(t) = \langle \omega_0, \kappa_t \rangle_{\mathfrak{H}}, \qquad (t \in E),$$

and

$$\|f\|_{R_k} = \|\omega_0\|_{\mathfrak{H}}.$$

The reproducing equation for R_k is

$$f(t) = \langle f, k(\cdot, t) \rangle. \tag{6.3}$$

It is often supposed that $\{\kappa_t\}$, $(t \in E)$ is complete in $\mathfrak{H}$, a rather mild restriction. This means that the only possible ω in (6.2) is ω_0, because from (6.1) the null space of $\mathcal{L}$ is $\{\theta\}$. Then $\mathcal{L} : \mathfrak{H} \mapsto R_k$ is an isometry and therefore bounded. It is clearly linear, one-to-one and 'onto'. Hence by the bounded inverse theorem, $\mathcal{L}^{-1}$ is bounded. These properties show that $\mathcal{L}$ is an isometric isomorphism of $\mathfrak{H}$ onto R_k.

6.1.3 Possibilities for extension

The r.k. theory outlined above is in some ways more general and in some ways less general than the LCA group setting. It is tied to Hilbert space methods and no extension beyond these methods is known to the authors.

However, in order to find some process of extension beyond Hilbert space there might be a way forward if a regime such as the following could be established. To describe such a regime we will adopt the following notations.

Suppose that $\mathfrak{X}$ is a linear space, and let $\mathfrak{X}^{\text{ex}}$ denote a linear space that extends $\mathfrak{X}$; that is, $\mathfrak{X}$ is a subspace of $\mathfrak{X}^{\text{ex}}$. Again, suppose that $\mathcal{P}$ denotes a proposition associated with $\mathfrak{X}$; it might be, for example, the assertion that a certain formula holds for all members of $\mathfrak{X}$. Let $\mathcal{P}^{\text{ex}}$ denote an extension of $\mathcal{P}$, that is, a proposition associated with $\mathfrak{X}^{\text{ex}}$ so chosen that it reduces to $\mathcal{P}$ for members of $\mathfrak{X}$.

With these notations, the following extension procedure can be suggested for the r.k. theory. Let $\mathfrak{H}^{\text{ex}}$ denote a linear space which extends $\mathfrak{H}$ in such a way as to extend the domain of $\mathcal{L}$, thus, $\mathcal{L}$ on this larger domain induces an extension R_k^{ex} of R_k. Let $\mathcal{P}$ denote a proposition associated with R_k, and $\mathcal{P}^{\text{ex}}$ a proposition associated with R_k^{ex}, asserting that $\mathcal{P}$ holds for R_k^{ex} but only in an approximate form when associated with $R_k^{\text{ex}} \backslash R_k$.

A scheme of this kind might afford some way of generalizing results found in the main body of the paper, such as the passage from the classical to the approximate sampling theorem (1.7)/(1.6), or the reproducing equation (1.8) to the approximate reproducing equation (1.9). But the construction of such a scheme is an open problem, no doubt a rather difficult one.

6.1.4 A dictionary

The question arises as to whether the ASF of (1.6) and ARKF of (1.9), introduced in this paper, as well as GPDF of (1.1) can be built into the more recent reproducing kernel Banach space theory (see e. g. [1,32,37,47,60,61]), since the Hilbert space approach may not suffice. Its *non-constructed* but practical applications would be precisely the present six theorems under discussion. Perhaps an extension of the Banach space theory may also be necessary. It is seemingly a wide open field (Table 1).

6.2 Abstract harmonic analysis

In abstract harmonic analysis, the real line in classical Fourier analysis is replaced by an LCA group. A very brief description is now given, following Rudin's notation and terminology in [76]. Further details can also be found in [4,25,38,39,73] and briefer

Table 1 A classical-reproducing kernel theory dictionary

Classical	r.k. theory
$\mathbb{R}$	E, an abstract set
$L^2(-\pi, \pi)$	$\mathfrak{H}$, a separable Hilbert space
$(2\pi)^{-1/2} e^{-it\cdot} \chi_{[-\pi,\pi]}(\cdot)$	$\kappa_t \in \mathfrak{H}, (t \in E)$
reproducing kernel: $\mathrm{sinc}(s-t)$	$k(s,t) = \langle \kappa_t, \kappa_s \rangle_{\mathfrak{H}}$
$\mathcal{F}^{-1}$, inverse Fourier transform	$\mathcal{L}$
$\hat{B}_\sigma^p$	R_k
reproducing equation $f(t) = \langle f, \mathrm{sinc}(\cdot - t) \rangle$	$f(t) = \langle f, k(\cdot, t) \rangle_{R_k}$

accounts are in [6,7,28]. The website [65] of the Numerical Harmonic Analysis Group (NuHAG) at the University of Vienna is a comprehensive and up-to-date source of information about all aspects of sampling theory.

6.2.1 Locally compact abelian groups

The property of being an LCA group is often preserved under processes that are subject to natural technical restrictions. Thus closed subgroups of an LCA group are also LCA, as are quotient groups of closed subgroups and so on. LCA groups enjoy a translation invariant measure m_G, called Haar measure, which is unique up to a multiplicative constant (for the real line, the Haar measure is the familiar Lebesgue measure). As with Lebesgue measure, Haar measure gives rise to an integral $\int_G f(x)dm_G(x) = \int_G f$; the notation $f \in L^p(G)$ is used if $\int_G |f|^p < \infty$, $p \geq 1$. As usual the statement that a property holds for (Haar) almost all points in a subset of G means that it holds for all points in the set except for a set of (Haar) measure 0.

6.2.2 The abstract Fourier transform

The exponential function $t \mapsto e^{iut}$ is replaced by a continuous homomorphism or *character* $\gamma\colon G \to \mathbb{S}^1$, which takes values on the unit circle. Under pointwise multiplication, these homomorphisms form an abelian group $G^\wedge := \Gamma$ which is locally compact with the compact-open topology. Greek letters such as γ and λ will be used for elements in Γ, which is called the *dual* group of G and corresponds to the frequency domain. The dual of Γ is isomorphic to G. This duality allows the value $\gamma(x), x \in G$, of a character γ to be written as an ordered pair (x, γ), where $(x, \gamma)(x', \gamma) = (x+x', \gamma)$ and similarly for γ.

The Fourier transform $\widehat{f}\colon \Gamma \to \mathbb{C}$ of the function $f \in L^1(G)$, defined by

$$\widehat{f}(\gamma) = \int_G f(x)(x, -\gamma)dm_G(x), \tag{6.4}$$

is continuous and vanishes at ∞. For each ψ in $L^1(\Gamma)$, the inverse Fourier transform-function $\psi^\vee\colon G \to \mathbb{C}$ is defined by

$$\psi^\vee(x) = \int_\Gamma \psi(\gamma)(x, \gamma)dm_\Gamma(\gamma). \tag{6.5}$$

The Haar measure m_Γ on Γ can be normalized so that the inversion formula

$$f(x) = \int_\Gamma \widehat{f}(\gamma)(x, \gamma)dm_\Gamma(\gamma) \tag{6.6}$$

holds almost always for suitable $f = \psi^\vee$ (see [76, Sect. 1.5] or [39, Thm. 31.17]).

These ideas are the basis of a beautiful and broad abstract analogue of classical Fourier analysis that retains the principal results of classical theory. Thus the abstract

Table 2 Some LCA groups, their duals, measures and characters

Group G	measure m_G	dual Γ	measure m_Γ	character (x, γ)
$\mathbb{R}$	Lebesgue	$\mathbb{R}$	Lebesgue	$e^{iv\xi}$
$\mathbb{R}^r$	Lebesgue	$\mathbb{R}^r$	Lebesgue	$e^{i\mathbf{v}\cdot\boldsymbol{\xi}}$
$\mathbb{Z}^s$	Point measure	$\mathbb{T}^s$	induced Lebesgue	$e^{i\mathbf{k}\cdot\boldsymbol{\xi}}$
$\mathbb{Z}_n^s$	Point measure	$\mathbb{Z}_n^s$	point measure	$e^{i\mathbf{k}\cdot\boldsymbol{v}}$

Table 3 A classical-abstract dictionary for sampling theory

Classical	Abstract
Time domain $\mathbb{R}$	LCA group G
Frequency domain $\mathbb{R}$	dual LCA group $G^\wedge = \Gamma$
Lebesgue measure $\lvert \cdot \rvert$	Haar measure m_G
e^{ixu}	character (x, γ)
$\widehat{f}(u) := \int_{\mathbb{R}} f(x)e^{-ixu}dx$	$\widehat{f}(\gamma) := \int_G f(x)(x, -\gamma)dm_G(x)$
$2w\mathbb{Z}$ discrete subgroup of $\mathbb{R}$	Λ discrete subgroup (lattice) of Γ
$\mathbb{R}/(2w\mathbb{Z}) \cong \mathbb{S}^1$	Γ/Λ compact abelian group
$(-\pi w, \pi w]$ transversal of $\mathbb{R}/(2\pi w\mathbb{Z})$	Ω transversal of Γ/Λ
$\mathbb{Z}/w = (2w\pi\mathbb{Z})^\perp$ sampling set	$H = \Lambda^\perp = \{h \in G : (h, \lambda) = 1, \lambda \in \Lambda\}$
$\lvert(-\pi w, \pi w]\rvert = 2\pi w$	$m_\Gamma(\Omega) < \infty$
$\sum_{k\in\mathbb{Z}} g(k/2w)$	$\int_H g(h)dm_H(h) = m_H(\{0\})\sum_{h\in H} g(h)$
$\widehat{B}_{\pi w}^p$ band-limited signals	$\widehat{B}^p(G)$
F^p includes non band-limited signals	$F^p(G)$
$\ell^p(\mathbb{Z}/w)$	$\ell^p(H)$

Fourier–Plancherel transform can be defined on $L^2(G)$ and analogues of the classical Fourier–Plancherel theorem ($\int_G \lvert f\rvert^2 = \int_\Gamma \lvert f^\wedge\rvert^2$) [39, Sect. 31], [76] and the Parseval theorem hold (here and where appropriate $\widehat{}$ is the Fourier–Plancherel transform). The abstract Poisson summation formula also holds under certain integrability conditions [76].

The general and unifying framework offered by abstract Fourier analysis includes n-dimensional Euclidean space as an important special case. Some other concrete examples of LCA groups, their duals, measures and characters are provided in Table 2; $\mathbb{Z}_n$ is the finite additive group of residues mod n and $\mathbb{T}^r = \mathbb{R}^r/\mathbb{Z}^r$ is the r-dimensional torus which we will take to be $[-1/2, 1/2]^r$, the r-fold product of the unit interval with endpoints identified. In particular, $\mathbb{T}^1 \cong \mathbb{S}^1$, the unit circle. This normalization differs slightly from that adopted in Table 3.

6.3 Abstract sampling theory

Sampling theory fits naturally into the abstract Fourier analysis setting, as the table above indicates, with different LCA groups giving rise to a variety of seemingly disparate sampling results [4]. Of course, there are some limitations to be expected in such a general theory. To explain it more fully, some additional definitions and notation are needed.

Let Λ be a *lattice in* Γ, i.e., a countable discrete subgroup of Γ with compact quotient Γ/Λ. Note that it is appropriate to work with the dual group Γ here, as it corresponds to the frequency domain. The lattice Λ corresponds to the sampling set in time domain.

The *annihilator*

$$\Lambda^\perp := H = \{h \in G \colon (h, \lambda) = 1 \text{ for all } \lambda \in \Lambda\}$$

of Λ is a closed subgroup which we will write H and which satisfies $H^\perp = \Lambda^{\perp\perp} = \Lambda$ [76, Lemma 2.1.3]. The annihilator H of Λ is isomorphic to the dual of Γ/Λ, i.e., $H \cong (\Gamma/\Lambda)^\wedge$ (algebraically and topologically) [76, Thm. 2.1.2] and is identified with the dual $(\Gamma/\Lambda)^\wedge$. The annihilator H plays the role of the sampling set in the case of the LCA group G. When $G = \Gamma = \mathbb{R}$ with lattice $\Lambda = 2\pi w\mathbb{Z}$, H reduces to the sampling set $\mathbb{Z}/w$ in (1.7).

The Haar measure m_Λ on Λ will be normalized so that the *Weil coset decomposition formula*

$$\int_\Gamma \varphi(\gamma)\, dm_\Gamma(\gamma) = \int_{\Gamma/\Lambda} \int_\Lambda \varphi(\gamma + \lambda)\, dm_\Lambda(\lambda)\, dm_{\Gamma/\Lambda}([\gamma]) \qquad (6.7)$$

holds for suitable $\varphi \colon \Gamma \to \mathbb{C}$ (see [39, Sect. 28.54 (iii)] or [73, Sect. 2.7.3]). The Haar measure $m_{\Gamma/\Lambda}$ of Γ/Λ is normalized so that the corresponding inversion formula (6.6) holds (with G replaced by Γ/Λ and Γ by H). The Haar measures of H and G/H are also normalized so that the Weil formula (6.7) holds for G and H and so that the corresponding inversion formula (6.6) holds (with G replaced by G/H and Γ by Λ).

The quotient group Γ/Λ has a complete set of coset representatives, also referred to as a *transversal*, the terminology we will use. Transversals are not unique and there is always a measurable one [29] which we will choose. Note that while the quotient group is compact, the transversal is only of finite measure; indeed unbounded transversals are possible (see [5, Sect. 4]). By definition, a transversal Ω consists of just one point from each distinct coset $[\gamma] = \Lambda + \gamma$, i.e., $\Omega \cap (\Lambda + \gamma)$ consists of a single point in Ω. Thus translates of Ω by non-zero elements in Λ are disjoint from Ω. Using (6.7), it is straightforward to verify that the Haar measure of the transversal Ω of Γ/Λ satisfies $m_\Gamma(\Omega) = m_\Lambda(\{0\})m_{\Gamma/\Lambda}(\Gamma/\Lambda) = m_\Lambda(\{0\})/m_H(\{0\}) < \infty$.

When the group G is discrete and countable, the integral over G reduces to a sum. This is particularly exploited for the subgroup Λ of Γ which is discrete and assumed countable, so that the integral

$$\int_\Lambda \varphi(\lambda)\, dm_\Lambda(\lambda) = m_\Lambda(\{0\}) \sum_{\lambda \in \Lambda} \varphi(\lambda),$$

where $m_\Lambda(\{0\})$ is the (discrete) measure of $\{0\}$.

6.3.1 Exact sampling in abstract harmonic analysis

Kluvánek [54] was the first to place sampling theory in the LCA group setting by
establishing the abstract analogue of the classical sampling theorem (i.e., the exact
sampling formula (1.7)). He considered LCA groups G, each with dual Γ having a
lattice Λ (so that Λ is a discrete, countable subgroup of Γ and Γ/Λ is compact). A
measurable transversal Ω, with necessarily finite Haar measure $m_\Gamma(\Omega)$, is taken to
represent the spectrum of f. Then Kluvánek took as an analogue of the class $\widehat{B}^2_{\pi w}$
(defined in Sect. 2.1 and called PW_w elsewhere) of band-limited functions, the class

$$\widehat{B}^2_\Omega(G) := \{f \in L^2(G) \cap C(G) \colon \operatorname{supp} \widehat{f} \subseteq \Omega\}. \tag{6.8}$$

Kluvánek showed that functions f in $\widehat{B}^2(G)$ have the representation $f = S_H f$,
where $S_H f$ is the abstract analogue of the classical sampling series $S_w f$ (2.18) and
is given by

$$(S_H f)(x) := \frac{1}{m_\Gamma(\Omega)} \sum_{h \in H} f(h)\chi_\Omega(x - k),$$

where convergence is in the L^2 sense and uniform. He further showed that $\|f\|^2_G = \int_G |f|^2 = \sum_{h \in H} |f(h)|^2$, the abstract sampling analogue for Plancherel's theorem,
equivalent to that for Parseval's theorem: $\int_G f\overline{g} = \sum_{h \in H} f(h)\overline{g}(h)$ and obviating
the need for the $\ell^2(H)$ summability condition.

The discrete Fourier transform, involving essentially $\mathbb{Z}$ and $\mathbb{S}^1$, also fits naturally
into this framework. The abstract analogue would involve a discrete abelian group and
a compact group respectively. A similar pattern holds for multi-dimensional signals
in higher dimensions.

6.3.2 Approximate sampling in abstract harmonic analysis

More recently, abstract analogues of the approximate or generalized sampling formula
(ASF) have been established. Faridani [28] proved a very general approximation theorem under an integrability condition. In [7,26], an abstract approximate sampling
theorem has been established under the square summability condition $f \in \ell^2(H)$
(in Sect. 2.1, $f \in S^2_w = \ell^2(\mathbb{Z}/w)$). The generalization of the classical sets F^p and
S^p_w, $p = 1, 2$ (defined in Sect. 2.1), which are appropriate for extending approximate
sampling theory to LCA groups, are

$$F^p(G) := \{f \in L^p(G) \cap C(G) \colon \widehat{f} \in L^1(\Gamma)\}$$

and

$$S^p_H(G) := \{f : G \to \mathbb{C} \colon f \in \ell^p(H)\} = \ell^p(H).$$

The abstract analogue of the ASF (1.6), is now given.

Theorem 3 *Suppose that the dual Γ of the locally compact abelian group G has a discrete subgroup Λ with Γ/Λ compact and that Ω is a measurable transversal of Γ/Λ. Then each $f \in F^2(G) \cap \ell^2(H)$ has a representation*

$$f = S_H f + R_H f,$$

where $H = \Lambda^\perp$ and $|(R_H f)(x)| \le 2 \int_{\Gamma \setminus \Omega} |\widehat{f}|$.

An asymptotic formula needs more information [7, Thms. 5, 6].

A norm result for f corresponding to Kluvánek's theorem is not known, although putting $f = g$ in the generalized or approximate Parseval formula (1.1) suggests that modulo some summability conditions

$$\int_G |f|^2 = \sum_{h \in H} |f(h)|^2 - E_1 + E_2, \qquad (6.9)$$

where

$$E_1 = \frac{1}{m_\Gamma(\Omega)} \sum_{h \in H} f(h) \int_{\Gamma \setminus \Omega} (\overline{f})^\wedge(\gamma)\,(h, \gamma)d\gamma, \;\; E_2 = \int_G (R_H f)\overline{f}.$$

This would imply an abstract Parseval type result and merits investigation. However, the proof in [18] of the formula (1.1) uses some subtle Fourier analysis similar to that employed by Brown [11, 12] and Boas [9] so that extra hypotheses would be very likely be needed.

Turning to the question of extending the equivalence results in Sect. 3, it could be conjectured possible that under suitable hypotheses, the abstract analogue of ASF (1.6) is equivalent to those of ARKF (1.9) and GPDF (1.1). However, the proof in [45] that the functional equation of the zeta function ζ is equivalent to the classical exact sampling theorem is far from direct, requiring long chains of implications and many results from complex analysis. Thus proving an LCA group result would appear to be very difficult, undoubtedly involving extra hypotheses. Moreover, even defining a suitable zeta function for LCA groups seems problematic as the natural numbers $\mathbb{N}$ form a multiplicative semigroup generated by primes. Nevertheless, the classical situation suggests some interesting lines of inquiry and is a useful guide to the further understanding of abstract Fourier analysis.

6.4 Sampling in Hilbert spaces, on Riemannian manifolds and graphs

Let L be a self-adjoint operator in a Hilbert space H and P_λ, $\lambda \in \mathbb{R}$, its spectral resolution of identity. With every vector f in H the family of projectors P_λ associates a measure on $\mathbb{R}$ which is given by the formula $\langle P_\lambda f, f \rangle$, where $\langle \cdot, \cdot \rangle$ is the inner product in H. Using this framework I. Pesenson introduced Paley–Wiener-type subspaces $PW_\omega(L)$, $\omega > 0$, by saying that f belongs to $PW_\omega(L)$ if and only if the measure $\langle P_\lambda f, f \rangle$ has support in $[-\omega, \omega]$. It turned out that these subspaces enjoy all the basic properties of the classical Paley–Wiener spaces and in particular they are suitable for development of a rich sampling and approximation theories. On the level of abstract Hilbert spaces it was done by Pesenson in [67,70]. Moreover, he specified this abstract setup in a number of very important situations such as: Riemannian compact and non-compact manifolds of bounded geometry [69], stratified Lie groups [68], combinatorial and quantum graphs [71,72].

7 A short biography of Wolfgang Splettstößer

Wolfgang Splettstößer, born in Düsseldorf on June 19, 1950, received his Abitur from the Leibniz Gymnasium in Düsseldorf in 1968, began studying mathematics at the RWTH Aachen in 1970, and received the Dipl. Math. degree with the bestowal of the "Springorum Denkmünze" in 1975. That year the electrical engineer Otto Lange (Aachen/TU Hamburg-Harburg) suggested one ought to examine in detail the broad area of the Shannon sampling theorem. Having attended the "1. Aachener Kollo-quium", devoted to "Special problems of signal theory", conducted by the late Hans Dieter Lüke that March, the senior author (PLB) realized that the topic was indeed an important one and Wolfgang Splettstößer would be just the right person for this field. Within a short time he performed pioneering research in sampling theory which led to his Dr. rer. nat. degree (with distinction) in 1977.

That same year the DFG-Schwerpunkt Program (priority programme) "Digitale Signalverarbeitung" was established; it consisted of a group of some 45 communica-tion engineers, geophysicists, seismologists and medical doctors which met ca. twice a year at various universities in order to discuss the research work achieved by their students. For the senior author, as a member of that group, the ideal student at his Lehrstuhl A für Mathematik was naturally Splettstößer. One of the many key prob-lems dealt with by this unique priority programme was prediction theory, brought up particularly by the late H. W. Schüßler (Erlangen), Alfred Fettweis (Bochum) and H. D. Lüke. Wolfgang Splettstößer decided to tackle prediction theory and already in December of 1981 he presented his Habilitation thesis to the RWTH Faculty. In view of his fundamental research work in signal analysis as a whole he was nominated Professor (apl.) at the Lehrstuhl A für Mathematik in 1987. His work was basic for a number of the ca. 140 papers, doctoral and diploma theses written by members of the chair in the area up until 1994. As to the status of sampling theory in the year 1975 the reader may consult [21, pp. 29–30].

His interest in applied mathematics, in particular in signal processing theory, led Wolfgang Splettstößer to carry on his career in industry, namely in the field of micro-electronic development, his first position in 1986 being a development engineer at Siemens Semiconductors, which became Infineon Technologies in 1999. He was pro-moted to Vice President of the Design-Center Düsseldorf in 1998, and finally took the role of the Senior Director of Human Resources at the Munich site in 2005. His deep concern for a dialogue-oriented management style motivated his continuous effort in the branch of "idea management", of which he became the global director of Infineon Technologies in 2007.

Furthermore, Wolfgang Splettstößer was director and member of the management board of the VDE (Association for Electrical, Electronic & Information Technologies)

Rhein Ruhr, from 2003 onwards. Some of his core interests, namely contacts between industry and universities, and a strong support of young researchers and engineers, for example as a selection-committee member for the "technics prize" awarded by the VDE Rhein-Ruhr, kept him connected with university during his whole career.

He passed away on March 18, 2013, and is survived by his spouse, Dr. Brigitte Splettstößer (née Koch), four daughters and a son.

Acknowledgments The authors would like to thank Professor Janine Splettstößer, Wolfgang's oldest daughter, for her contribution to the biography. Paulo Ferreira was partially supported by FCT under PTDC/EEA-TEL/108568/2008.

References

1. Aldroubi, A., Gröchenig, K.: Nonuniform sampling and reconstruction in shift-invariant spaces. SIAM Rev. **43**(4), 585–620 (2001).
2. Apostol, T.M.: Modular Functions and Dirichlet Series in Number Theory, 2nd edn. Springer, New York (1990).
3. Aronszajn, N.: Theory of reproducing kernels. Trans. Am. Math. Soc. **68**, 337–404 (1950)
4. Beaty, M.G., Dodson, M.M.: Abstract harmonic analysis and the sampling theorem. In: Higgins, J.R., Stens, R.L. (eds.) Sampling Theory in Fourier and Signal Analysis, vol 2: Advanced Topics, chap. 10, pp. 233–265. Oxford University Press, Oxford (1999)
5. Beaty, M.G., Dodson, M.M.: The Whittaker-Kotel'nikov-Shannon theorem, spectral translates and Plancherel's formula. J. Fourier Anal. Appl. **10**(2), 179–199 (2004).
6. Beaty, M.G., Dodson, M.M., Eveson, S.P.: A converse to Kluvánek's theorem. J. Fourier Anal. Appl. **13**(2), 187–196 (2007).
7. Beaty, M.G., Dodson, M.M., Eveson, S.P., Higgins, J.R.: On the approximate form of Kluvánek's theorem. J. Approx. Theory **160**(1–2), 281–303 (2009).
8. Bellman, R.: A Brief Introduction to Theta Functions. Holt, Rinehart and Winston, New York (1961)
9. Boas Jr, R.P.: Summation formulas and band-limited signals. Tôhoku Math. J. **2**(24), 121–125 (1972).

10. Bochner, S.: Vorlesungen über Fouriersche Integrale. Chelsea Publishing Co., New York (1948)
11. Brown Jr, J.L.: On the error in reconstructing a non-bandlimited function by means of the bandpass sampling theorem. J. Math. Anal. Appl. **18**(1), 75–84 (1967).
12. Brown Jr, J.L.: Erratum to: "On the error in reconstructing a nonbandlimited function by means of the bandpass sampling theorem". J. Math. Anal. Appl. **21**, 699 (1968).

13. Brüdern, J.: Einführung in die analytische Zahlentheorie. Springer, Berlin (1995)
14. Butzer, P.L., Schmeisser, G., Stens, R.L.: The classical and approximate sampling theorems and their equivalence for entire functions of exponential type. J. Approx. Theory **179**, 94–111 (2014).

15. Butzer, P.L., Ferreira, P.J.S.G., Higgins, J.R., Schmeisser, G., Stens, R.L.: The sampling theorem, Poisson's summation formula, general Parseval formula, reproducing kernel formula and the Paley-Wiener theorem for bandlimited signals - their interconnections. Appl. Anal. **90**(3–4), 431–461 (2011).

16. Butzer, P.L., Ferreira, P.J.S.G., Schmeisser, G., Stens, R.L.: The summation formulae of Euler–Maclaurin, Abel–Plana, Poisson, and their interconnections with the approximate sampling formula of signal analysis. Results Math. **59**(3–4), 359–400 (2011).
17. Butzer, P.L., Gessinger, A.: A decomposition theorem for Parseval's equation; Connections with uniform and nonuniform sampling. In: Bilinskis, I., Cain, G., Marvasti, F. (eds.) SampTA'95, pp. 100–107. Institute of Electronics and Computer Science, Riga (1995)

18. Butzer, P.L., Gessinger, A.: The approximate sampling theorem, Poisson's sum formula, a decomposition theorem for Parseval's equation and their interconnections. Ann. Numer. Math. **4**(1–4), 143–160 (1997)

19. Butzer, P.L., Jansche, S.: A direct approach to the Mellin transform. J. Fourier Anal. Appl. **3**(4), 325–376 (1997).

20. Butzer, P.L., Nessel, R.J.: Fourier Analysis and Approximation. Academic Press, New York; Birkhäuser Verlag, Basel (1971)

21. Butzer, P.L., Schmeisser, G., Stens, R.L.: An introduction to sampling analysis. In: Marvasti, F. (ed.) Nonuniform Sampling: Theory and Practice, pp. 17–121. Kluwer/Plenum, New York (2001).

22. Butzer, P.L., Splettstößer, W.: Approximation und Interpolation durch verallgemeinerte Abtastsummen. Forschungsberichte des Landes Nordrhein-Westfalen Nr. 2708. Westdeutscher, Opladen (1977)

23. Butzer, P.L., Stens, R.L.: The Euler–MacLaurin summation formula, the sampling theorem, and approximate integration over the real axis. Linear Algebra Appl. **52**(53), 141–155 (1983).

24. Butzer, P.L., Stens, R.L.: The Poisson summation formula, Whittaker's cardinal series and approximate integration. In: Ditzian, Z., Meir, A., Riemenschneider, S.D., Sharma, A. (eds.) Second Edmonton Conference on Approximation Theory, pp. 19–36. American Mathematical Society, Providence (1983)

25. Deitmar, A.: A First Course in Harmonic Analysis, 2nd edn. Springer, New York (2005)

26. Dodson, M.M.: Approximating signals in the abstract. Appl. Anal. **90**(3–4), 563–578 (2011).

27. Doetsch, G.: Summatorische Eigenschaften der Besselschen Funktionen und andere Funktionalrelationen, die mit der linearen Transformationsformel der Thetafunktion äquivalent sind. Compositio Math. **1**, 85–97 (1935)

28. Faridani, A.: A generalized sampling theorem for locally compact abelian groups. Math. Comp. **63**(207), 307–327 (1994).

29. Feldman, J., Greenleaf, F.P.: Existence of Borel transversals in groups. Pac. J. Math. **25**, 455–461 (1968)

30. Ferrar, W.L.: Summation formulae and their relation to Dirichlet's series. Compositio Math. **1**, 344–360 (1935).

31. Flajolet, P., Gourdon, X., Dumas, P.: Mellin transforms and asymptotics: harmonic sums. Theoret. Comput. Sci. **144**(1–2), 3–58 (1995).

32. García, A.G., Portal, A.: Sampling in reproducing kernel Banach spaces. Mediterr. J. Math. Published online: 28 Nov 2012.

33. Hamburger, H.: Über die Riemannsche Funktionalgleichung der ξ-Funktion. Math. Z. **10**(3–4), 240–254 (1921).

34. Hamburger, H.: Über die Riemannsche Funktionalgleichung der ζ-Funktion. Math. Z. **11**(3–4), 224–245 (1921).

35. Hamburger, H.: Über die Riemannsche Funktionalgleichung der ζ-Funktion. Math. Z. **13**(1), 283–311 (1922).

36. Hamburger, H.: Über einige Beziehungen, die mit der Funktionalgleichung der Riemannschen ζ-Funktion äquivalent sind. Math. Ann. **85**(1), 129–140 (1922).

37. Han, D., Nashed, M.Z., Sun, Q.: Sampling expansions in reproducing kernel Hilbert and Banach spaces. Numer. Funct. Anal. Optim. **30**(9–10), 971–987 (2009).

38. Hewitt, E., Ross, K.A.: Abstract Harmonic Analysis, vol. I: Structure of Topological Groups. Integration Theory, Group Representations. Springer, Berlin (1963)

39. Hewitt, E., Ross, K.A.: Abstract Harmonic Analysis, vol. II: Structure and Analysis for Compact Groups. Analysis on Locally Compact Abelian Groups. Springer, New York (1970)

40. Hewitt, E., Stromberg, K.: Real and Abstract Analysis. A Modern Treatment of the Theory of Functions of a Real Variable. Springer, Berlin (1965)

41. Higgins, J.R.: Sampling Theory in Fourier and Signal Analysis. Clarendon Press, Oxford (1996)

42. Higgins, J.R.: A sampling principle associated with Saitoh's fundamental theory of linear transformations. In: Saitoh, S., Hayashi, N., Yamamoto, M. (eds.) Analytic Extension Formulas and Their Applications (Collected papers 2nd Internat. Congress of the Internat. Society for Analysis, its Applications and Computation (ISAAC'99), Fukuoka, Japan, Aug. 16–21, 1999, and Research Meeting on Applications of Analytic Extensions, Kyoto, Japan, Jan. 11–13, 2000). Kluwer Academic Publishers, Dordrecht, pp. 73–86 (2001)

43. Higgins, J.R.: Two basic formulae of Euler and their equivalence to Tschakalov's sampling theorem. Sampl. Theory Signal Image Process. **2**(3), 259–270 (2003)
44. Higgins, J.R.: H. Hamburger's collection of five equivalent results, and some later developments, Technical Report, p. 31 (2007)
45. Higgins, J.R.: The Riemann zeta function and the sampling theorem. Sampl. Theory Signal Image Process. **8**(1), 1–12 (2009)
46. Higgins, J.R.: Paley–Wiener spaces and their reproducing formulae. In: Ruzhansky, M., Wirth, J. (eds.) Progress in Analysis and Its Applications (Proceedings of 7th Internat. ISAAC Congress, London, Jul. 13–18, 2009, pp. 273–279. World Scientific Publisher, Hackensack (2010).

47. Hong, Y.M., Kim, J.M., Kwon, K.H.: Sampling theory in abstract reproducing kernel Hilbert space. Sampl. Theory Signal Image Process. **6**(1), 109–121 (2007)
48. Ismail, M.E.H., Nashed, M.Z., Zayed, A.I., Ghaleb, A.F., eds. Mathematical Analysis, Wavelets, and Signal Processing, Proceedings of Internat. Conference on Mathematical Analysis and Signal Processing, Cairo, Jan. 3–9, American Mathematical Society, Providence, 1995, (1994).

49. Ivić, A.: The Riemann Zeta-Function. Wiley, New York (1985)
50. Kahane, J.P., Mandelbrojt, S.: Sur l'équation fonctionnelle de Riemann et la formule sommatoire de Poisson. Ann. Sci. École Norm. Sup. **3**(75), 57–80 (1958)
51. Karatsuba, A.A., Voronin, S.M.: The Riemann Zeta-Function. Walter de Gruyter & Co., Berlin (1992).

52. Klusch, D.: The sampling theorem, Dirichlet series and Bessel functions. Math. Nachr. **154**, 129–139 (1991).
53. Klusch, D.: The sampling theorem, Dirichlet series and Hankel transforms. J. Comput. Appl. Math. **44**(3), 261–273 (1992).
54. Kluvánek, I.: Sampling theorem in abstract harmonic analysis. Mat.-Fyz. Časopis Sloven. Akad. Vied **15**, 43–48 (1965)
55. Knopp, M., Robins, S.: Easy proofs of Riemann's functional equation for $\zeta(s)$ and of Lipschitz summation. Proc. Am. Math. Soc. **129**(7), 1915–1922 (2001).
56. Krazer, A.: Lehrbuch der Thetafunktionen. Chelsea Publishing Co., New York (1970)
57. Mordell, L.J.: Poisson's summation formula and the Riemann zeta function. J. Lond. Math. Soc. **S1–4**(4), 285 (1929).
58. Murty, M.R.: Problems in Analytic Number Theory, 2nd edn. Springer, New York (2008)
59. Nashed, M.Z.: Inverse problems, moment problems, signal processing: un menage a trois. In: Siddiqi, A.H., Singh, R.C., Manchanda, P. (eds.) Mathematics in Science and Technology (Proceedings of Satellite Conference of the Internat. Congress of Mathematicians, New Delhi, India, Aug. 14–17, 2010). World Scientific Publishers, Hackensack, pp. 2–19 (2011)
60. Nashed, M.Z., Sun, Q.: Sampling and reconstruction of signals in a reproducing kernel subspace of $L^p(\mathbb{R}^d)$. J. Funct. Anal. **258**(7), 2422–2452 (2010).
61. Nashed, M.Z., Sun, Q., Xian, J.: Convolution sampling and reconstruction of signals in a reproducing kernel subspace. Proc. Am. Math. Soc. **141**(6), 1995–2007 (2013).

62. Nashed, M.Z., Walter, G.G.: Reproducing kernel Hilbert spaces from sampling expansions. In: Ismail, M.E.H., Nashed, M.Z., Zayed, A.I., Ghaleb, A.F. (eds.) Mathematical Analysis, Wavelets, and Signal Processing, pp. 221–226 (1994).
63. Newman, D.J.: Analytic Number Theory. Springer, New York (1998)
64. Nikol'skiĭ, S.M.: Approximation of Functions of Several Variables and Imbedding Theorems. Springer, New York (1975)
65. NuHAG (Numerical Harmonic Analysis Group).
66. Patterson, S.J.: An Introduction to the Theory of the Riemann Zeta-Function. Cambridge University Press, Cambridge (1988).
67. Pesenson, I.: Best approximations in a space of the representation of a Lie group (russian). Dokl. Akad. Nauk SSSR 302(5), 1055–1058 (1988). Trans. Soviet Math. Dokl. **38**(2), 384–388 (1989)
68. Pesenson, I.: Sampling of Paley–Wiener functions on stratified groups. J. Fourier Anal. Appl. **4**(3), 271–281 (1998).
69. Pesenson, I.: A sampling theorem on homogeneous manifolds. Trans. Am. Math. Soc. **352**(9), 4257–4269 (2000).

70. Pesenson, I.: Sampling of band-limited vectors. J. Fourier Anal. Appl. **7**(1), 93–100 (2001).

71. Pesenson, I.: Polynomial splines and eigenvalue approximations on quantum graphs. J. Approx. Theory **135**(2), 203–220 (2005).

72. Pesenson, I.: Sampling in Paley–Wiener spaces on combinatorial graphs. Trans. Am. Math. Soc. **360**(10), 5603–5627 (2008).

73. Reiter, H.: Classical Harmonic Analysis and Locally Compact Groups. Clarendon Press, Oxford (1968)

74. Riemann, B.: Über die Anzahl der Primzahlen unter einer gegebenen Größe. Monatsber. Königl. Preuss. Akad. Wiss. Berlin Nov. 1859, 671–680. Reprinted. In: Riemann, B., Narasimhan, R. (eds.) Gesammelte mathematische Werke, wissenschaftlicher Nachlass und Nachträge, (Based on the edition by Weber, H., Dedekind, R.). Springer, Berlin (1990)

75. Rooney, P.G.: Another proof of the functional equation for the Riemann zeta function. J. Math. Anal. Appl. **185**(1), 223–228 (1994).

76. Rudin, W.: Fourier Analysis on Groups. Interscience Publishers, New York (1962)

77. Saitoh, S.: Integral Transforms. Reproducing Kernels and Their Applications. Longman, Harlow (1997)

78. Schuster, W.: Ein kurzer Beweis der Funktionalgleichung der Riemannschen Zetafunktion (a short proof of the functional equation of the riemann zeta function). Aequationes Math. **70**(1–2), 191–194 (2005).

79. Siegel, C.: Bemerkung zu einem Satz von Hamburger über die Funktionalgleichung der Riemannschen Zetafunktion. Math. Ann. **86**(3–4), 276–279 (1922).

80. Stenger, F.: Numerical Methods Based on sinc and Analytic Functions. Springer, New York (1993)

81. Stenger, F.: Sinc convolution-a tool for circumventing some limitations of classical signal processing. In: Ismail, M.E.H., Nashed, M.Z., Zayed, A.I., Ghaleb, A.F. (eds.) Mathematical Analysis, Wavelets, and Signal Processing, pp. 227–240 (1994).

82. Titchmarsh, E.C.: The Theory of the Riemann Zeta-Function, 2nd edn. Clarendon Press, New York (1986)

83. Weiss, P.: An estimate of the error arising from misapplication of the sampling theorem. Notices Am. Math. Soc. **10**, 351 (1963). Abstract No. 601-54

84. Zayed, A.I.: Advances in Shannon's Sampling Theory. CRC Press, Boca Raton (1993)

85. Zygmund, A.: Trigonometric Series. vols. I, II, 2nd edn., reprinted with corrections and some additions (two volumes bound as one) edn. Cambridge University Press, London (1968)

5

Applications of classical approximation theory to periodic basis function networks and computational harmonic analysis

**Hrushikesh N. Mhaskar · Paul Nevai ·
Eugene Shvarts**

Abstract In this paper, we describe a novel approach to classical approximation theory of periodic univariate and multivariate functions by trigonometric polynomials. While classical wisdom holds that such approximation is too sensitive to the lack of smoothness of the target functions at isolated points, our constructions show how to overcome this problem. We describe applications to approximation by periodic basis function networks, and indicate further research in the direction of Jacobi expansion and approximation on the Euclidean sphere. While the paper is mainly intended to be

Communicated by S.K. Jain.

The research of H.N. Mhaskar was supported, in part, by grant DMS-0908037 from the National Science Foundation and grant W911NF-09-1-0465 from the U.S. Army Research Office. The research of P. Nevai was supported by a KAU grant.

H. N. Mhaskar
Department of Mathematics, California Institute of Technology, Pasadena, CA 91125, USA
e-mail: hmhaska@gmail.com

H. N. Mhaskar
Institute of Mathematical Sciences, Claremont Graduate University, Claremont, CA 91711, USA

P. Nevai (✉)
King Abdulaziz University, Jeddah, Saudi Arabia
e-mail: paul@nevai.us

E. Shvarts
Shvarts Scientific Services, Pacoima, CA 91331, USA

E. Shvarts
Department of Mathematics, University of California at Davis, Davis, CA 95616, USA
e-mail: eugene.shvarts@gmail.com

a survey of our recent research in these directions, several results are proved for the first time here.

1 Introduction

Two of the major developments during the last quarter of a century are the use of radial basis function (RBF) networks in learning theory, and wavelet analysis in computational harmonic analysis. The subject of radial basis function networks is highly studied, with applications in various fields of mathematics, sciences, engineering, biology, learning theory, and so forth. A *MathSciNet* and a *Web of Science* search on January 10 of 2013 for "radial basis function*" showed 1,262 and 12,697 citations, respectively. A similar search with "wavelet*" revealed 10,842 and 65,271 citations, respectively. Therefore, we will make no effort even to attempt to survey the entire subject, and, instead, we will focus on certain facts which have impacted our own research in these areas.

Please note that, for the sake of clarity, the notation used in the introduction may not be the same during the rest of the paper.

A major theme in learning theory is to discover a functional relationship in given data of the form $\{(\mathbf{x}_k, y_k)\}_{k=1}^M$ where $\mathbf{x}_k$'s are vectors in a Euclidean space $\mathbb{R}^q$ for some $q \in \mathbb{N}$, and y_k's are the corresponding function values, usually corrupted with noise. The problem is to find a *target function* (or *model*) $f : \mathbb{R}^q \to \mathbb{R}$ so that $f(\mathbf{x}_k) = y_k$, $k = 1, 2, \ldots, M$ or at least, $f(\mathbf{x}_k) \approx y_k, k = 1, 2, \ldots, M$. The first requirement is usually imposed either when the data is insufficient, or when it is important to reproduce the data exactly in the model. For example, in image registration applications, one needs to preserve some "landmarks". On the other hand, if the data is plentiful, but noisy, one does not wish to reproduce it exactly, but wishes to "fit" a smooth model to the data.

In either case, the problem is clearly ill-posed—there are infinitely many models f which meet the requirements. Therefore, the usual method to obtain the model is first to choose a class $\mathcal{M}$ of desired models, and find the desired model f so as to minimize a *regularization functional* such as

$$\sum_{k=1}^M (g(\mathbf{x}_k) - y_k)^2 + \delta \|\mathcal{L}g\| \tag{1.1}$$

over all $g \in \mathcal{M}$, where $\mathcal{L}$ is a *penalty functional* (usually, a differential operator), $\| \circ \|$ is a suitable norm, and δ is the *regularization parameter*. The least square fit is obtained by setting $\delta = 0$; letting $\delta \to \infty$ is equivalent to minimizing $\|\mathcal{L}g\|$ subject to interpolatory conditions. A very classical example of this approach has $\mathbf{x}_k \in [-1, 1]$, $\mathcal{M}$ is the class of all twice continuously differentiable functions, $\| \circ \|$ is the L^2 norm on $\mathbb{R}$, $\mathcal{L}g = g''$, and we let $\delta \to \infty$. In this case, one recovers the cubic spline interpolant for the data. A technique well-known in image processing is the TV minimization, where $\mathbf{x}_k \in [0, 1]^2$, $\| \circ \|$ is the L^1 norm, and $\mathcal{L}g$ is the so-called total variation of g.

Motivated by the example of spline functions, regularization or smoothing interpolation is used with many other penalty functionals in learning theory. In view of the

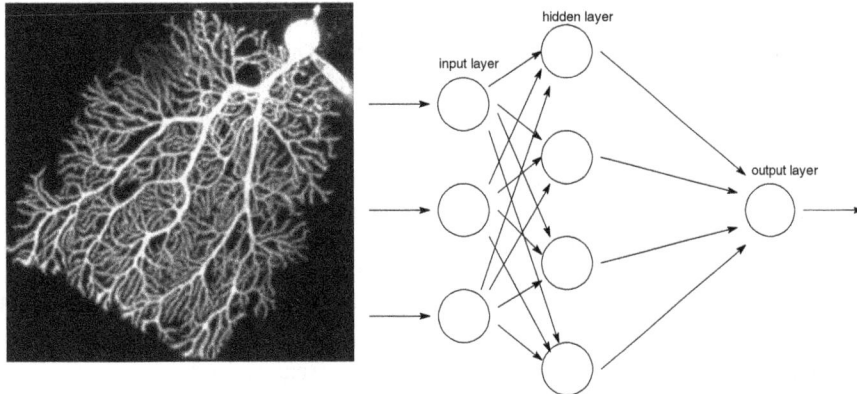

Fig. 1 The image on the left depicts a rat neuron. Neurons fire when their electric potential exceeds a certain threshold value. The term "neural network" arises in conjunction with the similarity found between the architecture of actual neuron cells, and the topology of an artificial neural RBF network, pictured on the *right*. The input layer receives a vector, and passes it as an argument to each of the small computers represented by *circles* in the hidden layer. Each of these computers evaluates a term of the form $G(\mathbf{x} - \mathbf{x}_j)$, and represents the analogous dependence on electric potential. The output layer takes the linear combination. The image copyrights are held by Testuya Tatsukawa 2010 and Unikom Center 2010–2012, respectively

Golomb–Weinberger principle, the solution to these problems can often be obtained explicitly in the form $\sum_{k=1}^{M} a_k \phi(|\circ -\mathbf{x}_k|)$ for a univariate function ϕ, where $|\circ|$ denotes the usual Euclidean norm [15,25]. A function of this form is called a *radial basis function* (RBF) *network* with M neurons. The function ϕ is called the *activation function* and the $\mathbf{x}_k$'s are the *centers* of the network. More generally, a *translation network* is a function of the form $\sum_{k=1}^{M} a_k G(\circ - \mathbf{x}_k)$ where the activation function G is defined on an appropriate Euclidean space.

The motivation for this terminology stems from an analogy with neurons in the brains, as indicated by Fig. 1.

A question of central interest in machine learning can be formulated as a problem of function approximation. A very classical theorem of Wiener states that the set of all translation networks is dense in $L^1(\mathbb{R}^q)$ if and only if $\hat{G}(\mathbf{t}) \neq 0$ for any $\mathbf{t} \in \mathbb{R}^q$. Under mild conditions on ϕ, it was proved by Park and Sandberg [70] that the class of all RBF networks

$$\left\{ \sum_{k=1}^{M} a_k \phi(|\circ -\mathbf{x}_k|) : M \geq 1, \ \mathbf{x}_k \in \mathbb{R}^q \right\}$$

is dense in the space of continuous real valued functions on $\mathbb{R}^q$ with the topology of convergence on compact subsets.

In [50], we showed that the set of all translation networks with activation function G which can be interpreted as a tempered distribution is dense in the same sense if and only if the support of the distributional Fourier transform of G is a so-called "set of uniqueness" for entire functions of finite exponential type in q variables.

Perhaps, the most popular way of using translation networks is for interpolating data. An attractive feature of RBF networks is a result by Micchelli [62] that (again under mild conditions on G) the matrix $[G(\mathbf{x}_j - \mathbf{x}_k)]_{j,k=1}^M$ is always invertible for an arbitrary choice of M and $\mathbf{x}_k$'s, so that interpolation by RBF networks is always possible. It is proved in [25] that an interpolatory network also solves certain regularization problems.

While interpolation by RBF networks as well as the use of extremal problems to obtain a functional relationship underlying a data are very popular methods, there are some disadvantages. First, there is no guarantee that as the data increases, the minimal value of the regularization functional will remain bounded. Second, there are usually no a priori bounds on the accuracy of the resulting approximation. Indeed, the theory of degree of approximation by interpolatory RBF networks is a fairly well established topic in its own right. Finally, there are numerous computational issues, including ill**-conditioning, lack of convergence, local minima, and so forth.

In the past 20 years or so, the first author, together with collaborators, has explored methods to construct translation network approximations with a priori performance guarantees which avoid each of the pitfalls mentioned above. Quite often, the resulting networks also satisfy the extremal properties up to a constant multiple. This research has demonstrated a very close connection between the classical theory of polynomial approximation, and approximation by translation networks. Applications of the ideas in this research include the theory of probability density estimation [78], pattern classification [3], control theory [33], signal processing [16], numerical simulation of turbulent channel flow [23], and construction of Schauder basis [24,37,31].

Some of the highlights of our work are the following:

- The conditions on the activation function required to achieve our approximation bounds are weaker than those required for interpolation.
- Our procedures are given as explicit formulas, requiring neither training in the classical machine learning setting, nor a solution of a system of equations involving a possibly ill-conditioned large matrix. The formulas can be implemented as a matrix–vector multiplication.
- The spaces to which the target function is assumed to belong are the usual smoothness classes, rather than the so-called native spaces for the networks.
- It is easy to adopt a two-scale approach to constructing a network which provides both the optimal approximation bounds, and interpolation at fewer points than there are centers of the network (cf. [7]).

The purpose of this paper is to illustrate the main ideas in our research in a few contexts. In Sect. 2, we introduce the basic ideas in the context of approximation by univariate trigonometric polynomials. The material in this section is extended in the context of multivariate trigonometric polynomials in Sect. 3. A new feature here is the ability to approximate functions based on "scattered data", that is, evaluations where one does not prescribe the location of the points where the function is to be evaluated. The analogues of these results in the context of periodic basis function networks are given in Sect. 4. Certain extensions of the various parts of this theory are discussed in Sect. 5.

2 Approximation of univariate periodic functions

2.1 Preliminaries

In this section we describe some basic results regarding approximation of univariate 2π-periodic functions by trigonometric polynomials. There are numerous standard references on the subject (for instance, [6,12,57,66,77]). Our summary in this section is based on [57], where many results are given with elementary proofs for the case of uniform approximation, and [12], where these results are given in the full generality. First, some terminology. We denote by $\mathbb{T}$ the quotient space of the interval $[-\pi, \pi]$ where the end points are identified. Geometrically, we think of $\mathbb{T}$ as the unit complex circle, except that rather than denoting a point on $\mathbb{T}$ by e^{ix}, we simplify the notation and denote it by x. If $f : \mathbb{T} \to \mathbb{R}$ is Lebesgue measurable, and $A \subseteq \mathbb{T}$ is Lebesgue measurable, we write

$$\|f\|_{p,A} \overset{\text{def}}{=} \begin{cases} \left\{ \frac{1}{2\pi} \int_A |f(t)|^p dt \right\}^{1/p}, & \text{if } 0 < p < \infty, \\ \text{ess sup}_{t \in A} |f(t)|, & \text{if } p = \infty. \end{cases} \quad (2.1)$$

The set of all Lebesgue measurable functions for which $\|f\|_{p,A} < \infty$ is denoted by $L^p(A)$, with the understanding that functions which are equal almost everywhere on A are considered equal as members of $L^p(A)$. The set of all uniformly continuous, bounded, and 2π-periodic functions on A, equipped with the norm of $L^\infty(A)$ (which is known as the *uniform* or *supremum* norm in this case) is denoted by $C^*(A)$. If $A = \mathbb{T}$, we will omit it from the notation. In the sequel, we will assume that $1 \le p \le \infty$. The *dual exponent notation* is as follows:

$$p' \overset{\text{def}}{=} \begin{cases} p/(p-1), & \text{if } 1 < p < \infty, \\ \infty, & \text{if } p = 1, \\ 1, & \text{if } p = \infty. \end{cases}$$

If n is a non-negative integer, then a *trigonometric polynomial of degree (or order)n* is defined to be a function of the form

$$x \mapsto \sum_{k=-n}^{n} c_k e^{ikx}, \quad x \in \mathbb{T},$$

where c_k's are complex numbers, known as the coefficients of the polynomial, and $|c_n| + |c_{-n}| \ne 0$. The set of all trigonometric polynomials of degree $< n$ will be denoted by $\mathbb{H}_n$. It is convenient to extend this notation to non-integer values of n by setting $\mathbb{H}_n \overset{\text{def}}{=} \mathbb{H}_{\lfloor n \rfloor}$ where $\lfloor n \rfloor$ is the integer part of n, and setting $\mathbb{H}_n = \{0\}$ if $n \le 0$. According to the trigonometric variant of the Weierstrass theorem and its L^p-versions, the L^p-closure of $\cup_{n=1}^{\infty} \mathbb{H}_n$ is L^p for $1 \le p < \infty$ and C^* for $p = \infty$. In order to avoid making an elaborate distinction between the two cases every time we state a theorem, we will write $X^p = L^p$ if $p < \infty$ and $X^\infty \overset{\text{def}}{=} C^*$. A central theme of approximation theory in this context is to investigate the properties of the *degree of approximation* of $f \in X^p$ from $\mathbb{H}_n$ for all n. This is defined for $f \in X^p$ and $n \geq 0$ by

$$E_{n,p}(f) \stackrel{\text{def}}{=} \inf\{\|f - P\|_p : P \in \mathbb{H}_n\}.$$

The quantity $E_{n,p}(f)$ measures the minimal error that one **must** expect if one wishes to use an element of $\mathbb{H}_n$ as a model for f. Clearly, the function $n \mapsto E_{n,p}(f)$ is non-increasing, the function $p \mapsto E_{n,p}(f)$ is non-decreasing, and the function $f \mapsto E_{n,p}(f)$ is a semi-norm on X^p if $p \geq 1$. In the sequel, we will assume that $1 \leq p \leq \infty$.

Two of the main problems in this theory are the following. One of them is how to construct the *best approximation* $P_p^*(f) \in \mathbb{H}_n$ such that $\|f - P_p^*(f)\|_p = E_{n,p}(f)$. The other problem is to give, for $\gamma > 0$, a complete characterization of $f \in X^p$ for which $E_{n,p}(f) = \mathcal{O}(n^{-\gamma})$. Except in the case when $p = 2$, the operator P_p^* is non-linear, and it takes a very elaborate optimization technique to compute $P_p^*(f)$ in general. Thus we adopt the stance that it is computationally worthwhile to construct *near-best approximations* or *sub-optimal solutions*, that is, finding some $P \in \mathbb{H}_n$ for which $\|f - P\|_p \leq c_1 E_{cn,p}(f)$ for some positive constants c, c_1 independent of n or f. We discuss these constructions in this subsection, postponing the discussion of the characterization question to Sect. 2.2.

2.1.1 Constant convention

In the remainder of this paper, the symbols $c, c_1, \ldots,$ will denote generic positive constants whose value is independent of the target function f, and other variable parameters such as n, but may depend on fixed parameters under discussion, such as the norm p or the smoothness index γ, and so forth. Their value may be different at different occurrences, even within a single formula. The notation $A \sim B$ will mean $c_1 A \leq B \leq c_2 A$. As usual, the notation $A = \mathcal{O}(B)$ will mean that $|A| \leq c|B|$, where c is some positive constant whose value may depend on f which one way or another may appear in the expressions A and B.

In the case when $p = 2$, an explicit formula for the best approximation polynomial $P_2^*(f)$ is well known. We define the *Fourier coefficients* of f by

$$\hat{f}(k) = \frac{1}{2\pi} \int_{\mathbb{T}} f(t) \exp(-ikt) \, dt, \quad k \in \mathbb{Z},$$

and for $n \in \mathbb{N}$ we set

$$s_n(f, x) = \sum_{|k| \leq n-1} \hat{f}(k) \exp(ikx), \quad x \in \mathbb{T}.$$

It is well known that $P_2^*(f) = s_n(f)$, that is,

$$\|s_n(f) - f\|_2 = \inf\{\|f - P\|_2 : P \in \mathbb{H}_n\}.$$

It is well known [79, Chapter VII, Theorem 6.4] that

$$\|s_n(f) - f\|_p \le cE_{n,p}(f), \quad 1 < p < \infty,$$

where c is a constant depending on p. The value of c tends to ∞ if $p \to 1, \infty$. There exist integrable functions f for which the sequence $\{s_n(f, x)\}$ diverges for almost all x, and a dense set of functions $f \in C^*$ for which $\{s_n(f, 0)\}$ diverges. A very deep theorem in the theory of Fourier series, the Carleson–Hunt theorem [4, 30] states that if $p > 1$ and $f \in L^p$ then the sequence $\{s_n(f, x)\}$ converges for almost all x to $f(x)$.

A ground-breaking theorem in this direction states that if

$$\sigma_n(f, x) \stackrel{\text{def}}{=} \frac{1}{n} \sum_{m=1}^{n} s_m(f, x), \quad n \in \mathbb{N} \text{ and } x \in \mathbb{T},$$

then for all p with $1 \le p \le \infty$ and $f \in X^p$,

$$\lim_{n \to \infty} \|f - \sigma_n(f)\|_p = 0.$$

In the case when $p = \infty$, this theorem was proved by Fejér in [18] and it appears in almost every textbook on approximation theory, see e.g., [12, Chapter 1, Corollary 2.2] or [57, Chapter 1, Section 1.1]. In the case of $1 \le p < \infty$ and, in fact, in greater generality, it appears in [79, Chapter IV, Theorem 5.14]. The rate of convergence

$$\|f - \sigma_n(f)\|_p \le \frac{c}{n} \sum_{k=1}^{n} E_{k,p}(f),$$

was given in 1961 by Stechkin (aka Stečkin), see [74]. In order to get a near-best approximation, we define

$$v_n(f, x) \stackrel{\text{def}}{=} \frac{1}{n} \sum_{m=n+1}^{2n} s_m(f, x), \quad n \in \mathbb{N} \text{ and } x \in \mathbb{T},$$

where v is in honor of C. de La Vallée Poussin. It is easy to check that

$$v_n(f) = 2\sigma_{2n}(f) - \sigma_n(f).$$

It can be deduced from here ([12, Chapter 9, Theorem 3.1]) that for all p such that $1 \le p \le \infty$ and for all $f \in L^p$ one has

$$E_{2n,p}(f) \le \|f - v_n(f)\|_p \le 4E_{n,p}(f).$$

To understand the difference between the behavior of $\sigma_n(f)$ and $v_n(f)$, we point out an alternative expression for these. Namely,

$$\sigma_n(f, x) = \sum_{|k|<n} \left(1 - \frac{|k|}{n}\right) \hat{f}(k) \exp(ikx),$$

$$v_n(f, x) = \sum_{|k|\leq n} \hat{f}(k) \exp(ikx) + \sum_{n+1\leq|k|<2n} \left(2 - \frac{|k|}{n}\right) \hat{f}(k) \exp(ikx). \quad (2.2)$$

Thus, if $P \in \mathbb{H}_n$ and P is not a constant, then $\sigma_n(P) \neq P$ but $v_n(P) \equiv P$. We observe further that $v_n(f)$ can be written in the form

$$v_n(f, x) = \sum_{|k|<2n} h\left(\frac{k}{2n}\right) \hat{f}(k) \exp(ikx),$$

where

$$h(t) = \begin{cases} 1, & \text{if } |t| \leq 1/2, \\ 2 - 2t, & \text{if } 1/2 < |t| < 1, \\ 0, & \text{if } |t| \geq 1. \end{cases}$$

Motivated by this observation, we make the following definition.

Definition 2.1 Let $h : \mathbb{R} \to [0, 1]$ be compactly supported. The **summability operator** (corresponding to the **filter** h) is defined for $f \in L^1$ by

$$\sigma_n(h, f, x) \stackrel{\text{def}}{=} \sum_{k\in\mathbb{Z}} h\left(\frac{k}{n}\right) \hat{f}(k) \exp(ikx), \quad n \in \mathbb{N} \,\&\, x \in \mathbb{T}. \quad (2.3)$$

The **summability kernel** corresponding to h is defined by

$$\Phi_n(h, t) \stackrel{\text{def}}{=} \sum_{k\in\mathbb{Z}} h\left(\frac{k}{n}\right) \exp(ikt), \quad t \in \mathbb{T}. \quad (2.4)$$

The function h is called a **low pass filter** if h is an even function, non-increasing on $[0, \infty)$, $h(t) = 1$ if $|t| \leq 1/2$, and $h(t) = 0$ if $|t| \geq 1$.

We note that for $n \in \mathbb{N}$ and $f \in L^1$,

$$\sigma_n(h, f, x) = \frac{1}{2\pi} \int_{\mathbb{T}} f(t)\Phi_n(h, x - t)\, dt$$

$$= \frac{1}{2\pi} \int_{\mathbb{T}} f(x - t)\Phi_n(h, t)\, dt, \quad f \in L^1, \, x \in \mathbb{T}. \quad (2.5)$$

Even though the sum in (2.3) is written as an infinite sum, it is in fact a finite sum since $h(|k|/n) = 0$ if $|k|$ is sufficiently large; if h is a low pass filter, then $k \geq n$ is sufficient. Therefore, for such h, $\sigma_n(h, f) \in \mathbb{H}_n$. By writing the sum as an infinite sum, we avoid the need to restrict ourselves in the definition to the case when n is an integer. If h is supported on $[-A, A]$, then $\Phi_{1/A}(h, t) \equiv 1$, and we redefine $\Phi_n(h, t) \equiv 0$ if $n < 1/A$.

We summarize some important properties of the summability operator. In what follows, the constants may depend on the function h.

Theorem 2.1 (a) *If $S \geq 2$ is an integer, and h is an S-times continuously differentiable function, then*

$$|\Phi_n(h, t)| \leq cn \min\left(1, (n|t|)^{-S}\right), \quad t \in \mathbb{T} \ \& \ n \geq 1. \tag{2.6}$$

(b) *Let h be a twice continuously differentiable, even function, $1 \leq p \leq \infty$. Then*

$$\|\sigma_n(h, f)\|_p \leq c\|f\|_p, \quad f \in L^p \ \& \ n \in \mathbb{N}. \tag{2.7}$$

(c) *Let h be a twice continuously differentiable low pass filter, $n \in \mathbb{N}$. Then $\sigma_n(h, P) \equiv P$ for all $P \in \mathbb{H}_{n/2}$. In addition, for $1 \leq p \leq \infty$ and $f \in L^p$,*

$$E_{n,p}(f) \leq \|f - \sigma_n(h, f)\|_p \leq cE_{n/2,p}(f). \tag{2.8}$$

Since the localization estimate (2.6) and its analogues in various contexts play an important role in our theory, we pause in our discussion to illustrate it by an example. We consider two low pass filters, h_3 and h_∞ defined on $(1/2, 1)$ by

$$h_3(t) = (1 - t)^3 \left(8 + 48(t - 1/2) + 192(t - 1/2)^2\right),$$

and

$$h_\infty(t) = \exp\left(-\frac{\exp(2/(1 - 2t))}{1 - t}\right).$$

Of course, both functions are equal to 1 on $[0, 1/2]$ and 0 on $[1, \infty)$. The function h_3 is twice continuously differentiable on $[0, \infty)$ whereas h_∞ is infinitely many times differentiable. In Fig. 2, we show the graphs of $|\Phi_n(h_3, x)|$ and $|\Phi_n(h_\infty, x)|$ for $n = 512$ and $n = 1024$ on the interval $[\pi/3, \pi]$. It is clear from the figure that for both the values of n, the graph corresponding to h_∞ is an order of magnitude smaller than that for h_3, and that the graph corresponding to h_∞ decreases much faster as n (and/or $|x|$) increases.

Proof of Theorem 2.1 Part (a) can be proved using the Poisson summation formula; see, e.g., [75, Chapter VII, Theorem 2.4 and Corollary 2.6]; a recent proof is given in

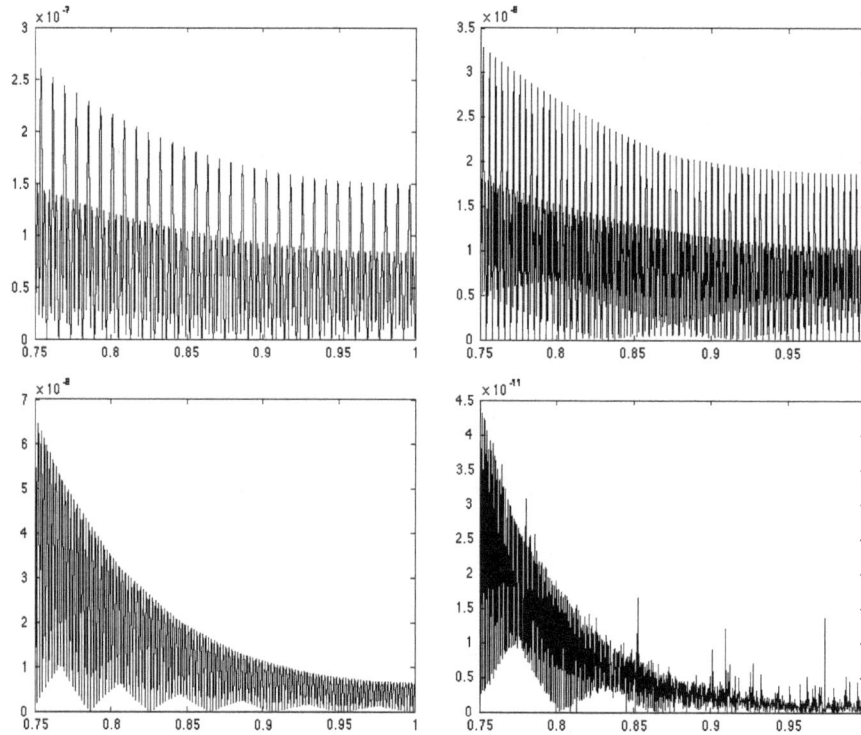

Fig. 2 Clockwise from the top left, the graphs of $|\Phi_{512}(h_3, x)|$, $|\Phi_{1024}(h_3, x)|$, $|\Phi_{1024}(h_\infty, x)|$, and $|\Phi_{512}(h_\infty, x)|$. The numbers on the x axis are in multiples of π. The maximum absolute values of these are $2.6002e-07$, $3.2776e-08$, $4.4293e-11$, and $6.6296e-08$ respectively

[22]. If h is twice continuously differentiable, we use (2.6) with $S = 2$ to obtain

$$
\int_{\mathbb{T}} |\Phi_n(h, t)| \, dt = \int_{|t| \leq 1/n} |\Phi_n(h, t)| \, dt + \int_{t \in \mathbb{T}, \, |t| > 1/n} |\Phi_n(h, t)| \, dt
$$

$$
\leq cn(2/n) + 2cn^{-1} \int_{1/n}^{\pi} t^{-2} \, dt \leq 2c + 2cn^{-1} \int_{1/n}^{\infty} t^{-2} \, dt = 4c.
$$

The estimate (2.7) is now deduced using the convolution identity (2.5) and Young's inequality (cf. [79, Chapter II, Theorem 1.15]). This proves part (b). Since h is a low pass filter, $h(|k|/n) = 1$ if $|k| < n/2$. If $P \in \mathbb{H}_{n/2}$ then $\hat{P}(k) = 0$ if $|k| \geq n/2$. Hence,

$$
\sigma_n(h, P, x) = \sum_{|k| < n/2} h\left(\frac{|k|}{n}\right) \hat{P}(k) \exp(ikx) = P(x), \quad x \in \mathbb{T}.
$$

This proves the first assertion in part (c). Let $f \in L^p$. The first inequality in (2.8) follows from the fact that $\sigma_n(h, f) \in \mathbb{H}_n$. If $P \in \mathbb{H}_{n/2}$ is arbitrary, then

$$\|f - \sigma_n(h, f)\|_p = \|f - P - \sigma_n(h, f - P)\|_p \le \|f - P\|_p + \|\sigma_n(h, f - P)\|_p$$
$$\le c_1\|f - P\|_p.$$

The second inequality in (2.8) follows by taking the infimum over $P \in \mathbb{H}_{n/2}$. $\quad\square$

2.2 Degree of approximation

If $\epsilon_n \downarrow 0$ as $n \to \infty$, then it is easy to construct $f \in C^*$ for which $E_{n,\infty}(f) \le \epsilon_n$. We set $\delta_k = \epsilon_k - \epsilon_{k+1}$ if $k \ge 0$, and observe that $\sum \delta_k$ is a convergent series with non-negative terms. Since $|\cos(kx)| \le 1$ for all $x \in \mathbb{T}$, the series

$$f(x) = \sum_{k=0}^{\infty} \delta_k \cos(kx)$$

converges uniformly and absolutely to $f \in C^*$. Clearly,

$$E_{n,\infty}(f) \le \sum_{k=n}^{\infty} \delta_k = \epsilon_n.$$

As a matter of fact, as proved by Bernstein (cf. [66, Chapter V, Section 5, p. 109]), one could even have $E_{n,\infty}(f) = \epsilon_n$. A central question in approximation theory is to determine the constructive properties (smoothness) of f which are equivalent to a given rate of decrease of the sequence $\{E_{n,p}(f)\}$. In particular, we will examine the very classical case where $E_{n,p}(f) = \mathcal{O}(n^{-\gamma})$ for some $\gamma > 0$.

The fundamental inequalities in this theory are given in the following theorem (for the uniform norm and elementary proofs, see [57, Chapter III, Section 3.1, Theorem 2, Chapter III, Section 2.2, Corollary 3] and in greater generality [12, Chapter 4, Theorem 2.4, Chapter 7, Section 4]). The best constant in the Bernstein inequality, see (2.9) below, is well-known. The idea of using summability operators to prove Bernstein-type inequalities goes back to 1914 when Riesz realized that Landau's trick of applying Fejér's ideas works for Bernstein-type inequalities. Riesz even managed to find the best constant with this seemingly inefficient approach; for the full story see [69]. While Riesz used an explicit expression for the kernel, we will use only the filter which gives rise to the summability kernel. We do not get the best constants, but the method can be generalized considerably to the settings when such an explicit expression for the kernel is not available, see, e.g., [39]. The best constants in (2.10) below are known [12, Chapter 7, Theorem 4.3], [57, Chapter III, Section 2.2, Theorem 4] as well.

Theorem 2.2 *Let $1 \le p \le \infty$, $n \in \mathbb{N}$, and let $r \in \mathbb{N}$.*

(a) *If $P \in \mathbb{H}_n$ then the Bernstein-type inequality*

$$\|P^{(r)}\|_p \le cn^r\|P\|_p \tag{2.9}$$

holds.

(b) *If $f \in X^p$ and $f^{(r)} \in X^p$, then the Favard-type inequality*

$$E_{n,p}(f) \leq \frac{c}{n^r} \| f^{(r)} \|_p. \tag{2.10}$$

holds.

Proof Let us fix an infinitely differentiable low pass filter h. Clearly, the function $h_1(t) \stackrel{\text{def}}{=} t^r h(t)$, $t \in \mathbb{R}$, is compactly supported and it is infinitely differentiable. To prove part (a), let $P \in \mathbb{H}_n$. Then, in view of Theorem 2.1(c), we have

$$P^{(r)}(x) = \sum_{k \in \mathbb{Z}} h\left(\frac{k}{2n}\right) \widehat{P^{(r)}}(k) \exp(ikx) = \sum_{k \in \mathbb{Z}} h\left(\frac{k}{2n}\right) (ik)^r \hat{P}(k) \exp(ikx)$$

$$= (2in)^r \sum_{k \in \mathbb{Z}} h_1\left(\frac{k}{2n}\right) \hat{P}(k) \exp(ikx) = (2in)^r \sigma_{2n}(h_1, P, x).$$

The estimate (2.9) now follows immediately from (2.7). This proves part (a). Next, we prove part (b). For $f \in X^p$, we define

$$P_j \stackrel{\text{def}}{=} \begin{cases} \sigma_1(h, f), & \text{if } j = 0, \\ \sigma_{2^j}(h, f) - \sigma_{2^{j-1}}(h, f), & \text{if } j \in \mathbb{N}. \end{cases} \tag{2.11}$$

Then for all integer $m \geq 2$,

$$\sum_{j=0}^{m} P_j = \sigma_{2^m}(h, f), \tag{2.12}$$

and so, since $f \in X^p$, (2.8) implies that

$$f = \sum_{j=0}^{\infty} P_j, \tag{2.13}$$

with the series converging in the sense of X^p.

Next, let $g(t) = h(t) - h(2t)$, and $g_1(t) = g(t)t^{-r}$. Then $g(t) = 0$ if $|t| \leq 1/4$, and, hence, g_1 is infinitely differentiable. For $j \in \mathbb{N}$ and $x \in \mathbb{T}$, we have

$$P_j(x) = \sum_{k \in \mathbb{Z}} \left(h\left(\frac{|k|}{2^j}\right) - h\left(\frac{|k|}{2^{j-1}}\right) \right) \hat{f}(k) \exp(ikx)$$

$$= \sum_{k \in \mathbb{Z}} g\left(\frac{|k|}{2^j}\right) \hat{f}(k) \exp(ikx) = \sigma_j(g, f, x).$$

Let $j \geq 2$. Taking into consideration that $g(t) = 0$ if $|t| \leq 1/4$, we can deduce that

$$
\begin{aligned}
P_j(x) = \sigma_{2^j}(g, f, x) &= \sum_{k \in \mathbb{Z}} g\left(\frac{k}{2^j}\right) \hat{f}(k) \exp(ikx) \\
&= \sum_{k \in \mathbb{Z}} g\left(\frac{k}{2^j}\right) \frac{1}{(ik)^r} \widehat{f^{(r)}}(k) \exp(ikx) \\
&= \frac{1}{(i2^j)^r} \sum_{k \in \mathbb{Z}} g_1\left(\frac{k}{2^j}\right) \widehat{f^{(r)}}(k) \exp(ikx) \\
&= \frac{1}{(i2^j)^r} \sigma_{2^j}(g_1, f^{(r)}, x).
\end{aligned}
$$

Therefore, (2.7) shows that

$$
\|P_j\|_p \leq c 2^{-jr} \|f^{(r)}\|_p.
$$

Using (2.12) and (2.13), we conclude that

$$
E_{2^m, p}(f) \leq \|f - \sigma_{2^m}(h, f)\|_p \leq \sum_{j=m+1}^{\infty} \|P_j\|_p \leq c \|f^{(r)}\|_p \sum_{j=m+1}^{\infty} 2^{-jr} = c_1 2^{-mr} \|f^{(r)}\|_p.
$$

If $n \geq 4$, we find $m \geq 2$ such that $2^m \leq n \leq 2^{m+1}$. Then the above estimate shows (2.10) in the case when $n \geq 4$. The estimate is trivial if $n = 1, 2, 3$. $\quad\square$

We pause in our discussion to introduce the notions of widths in approximation theory, in contradistinction with the characterization theorem (Theorem 2.3 below), which is our main interest in this section.

Let X be a normed linear space, $n \geq 1$ be an integer, $K \subset X$ be a compact set, and $Y \subset X$ be a closed set. We define the distance

$$
\operatorname{dist}(K, Y) \overset{\text{def}}{=} \sup_{f \in K} \inf_{P \in Y} \|f - P\|_X. \tag{2.14}
$$

In this paper, we will refer to bounds on $\operatorname{dist}(K, Y)$ as *distance bounds*.

The notion of widths in approximation theory gives in some sense minimal distance of K from various different choices of Y. Depending upon the nature of Y, there are different variants of widths [36,72]. We will describe two of these.

For integer $n \geq 1$, the *Kolmogorov n-width* of K [32] is defined by

$$
d_{n,kol}(K, X) \overset{\text{def}}{=} \inf_{X_n} \operatorname{dist}(K, X_n), \tag{2.15}
$$

where the infimum is taken over all linear subspaces X_n of X, with the dimension of X_n being $\leq n$. More generally, any process of approximation of elements of K based on n parameters can be formalized as the composition of two functions: the first, $\mathcal{M} : K \rightarrow \mathbb{R}^n$, represents the selection of the parameters, while the second, $\mathcal{R} : \mathbb{R}^n \rightarrow X$, represents the reconstruction. Their composition $f \mapsto \mathcal{R}(\mathcal{M}(f))$ is the desired approximation of $f \in K$. The nonlinear n-width of K with respect to X is defined by DeVore et al. [11] and independently by Mathé [41] as

$$d_n(K, X) \stackrel{\text{def}}{=} \inf \sup_{f \in K} \| f - \mathcal{R}(\mathcal{M}(f)) \|_X, \qquad (2.16)$$

where the infimum is taken over *all* $\mathcal{R} : \mathbb{R}^n \rightarrow X$, and all *continuous* $\mathcal{M} : K \rightarrow \mathbb{R}^n$. Here, the reconstruction operation can be arbitrary. The nonlinear n-width gives the theoretically minimum "worst-case error" in approximating elements of K, subject only to the prior knowledge that they are elements of K, using n parameters selected in a stable manner.

It is not immediately clear that the choice of parameters involved in approximation from an arbitrary finite dimensional space is always continuous. Therefore, it is not obvious that the nonlinear n-width is a sharpening of the Kolmogorov n-width. However, in view of [11, Corollary 2.2], it follows that

$$d_n(K, X) \leq d_{n,kol}(K, X).$$

In many applications, K is defined in terms of a semi-norm $\| \circ \|_K$ on a subset of X:

$$K = \{ f \in X : \| f \|_K \leq 1 \}.$$

As a consequence of [11, Theorem 3.1], it follows that if there exists a linear subspace $Y \subset X$ with dimension $n + 1$ such that a Bernstein inequality of the form

$$\| g \|_K \leq b_n(K) \| g \|_X, \quad g \in Y, \qquad (2.17)$$

holds, then with an absolute constant c,

$$d_n(K, x) \geq c b_n(K)^{-1}. \qquad (2.18)$$

For example, let $X = X^p$, $Y = \mathbb{H}_n$. Then $\text{dist}(K, \mathbb{H}_n)$ is the "worst-case error" in approximating elements of K by trigonometric polynomials of degree $< n$. When

$$K \stackrel{\text{def}}{=} \{ f \in X^p : \| f^{(r)} \|_p \leq 1 \}, \qquad (2.19)$$

the Favard inequality (2.10) shows the upper distance bound

$$\text{dist } (K, \mathbb{H}_n) \leq c n^{-r}.$$

Therefore, with $n = 2m - 1$, we see that $d_n(K) \leq d_{n,kol}(K) \leq c n^{-r}$, while the Bernstein inequality (2.9) and (2.18) together show that $d_n(K) \geq c n^{-r}$. Moreover, Theorem 2.1 then leads us to conclude that using the Fourier coefficients for the parameter selection, followed by the operator σ_n is, up to a constant factor, an optimal reconstruction algorithm for approximation of functions in K.

The estimates on the n-widths or the lower distance bounds imply that the degree of approximation or method of approximation cannot be improved, in the sense that there is always some "bad function" in the class for which a lower bound is attained. They do not address the question of whether *individual functions* can be approximated better than what the degree of approximation theorem predicts, based on the a priori information known about the target function. For example, one could conceivably use some clever ideas appropriate for the target function, that can yield a better performance than the theoretically assumed a priori information. This is the question of *characterization of smoothness classes*. This is our main interest in this section, and we now resume this discussion.

It is perhaps clear that the number of derivatives is not a sufficiently sophisticated indication to characterize functions for which $E_{n,p}(f) = \mathcal{O}(n^{-\gamma})$; e.g., when γ is not an integer. If γ is an integer and $f^{(\gamma)} \in X^p$, then $E_{n,p}(f) = \mathcal{O}(n^{-\gamma})$ as proved in (2.10). The converse is not true. For example, if $f(x) = |\cos x|$, then it is easy to compute that

$$f(x) = \frac{2}{\pi} + \frac{4}{\pi} \sum_{k=1}^{\infty} (-1)^{k+1} \frac{\cos(2kx)}{4k^2 - 1}$$

where the series converges uniformly. Therefore,

$$E_{2n-1,\infty}(f) \leq \|f - s_{2n}(f)\|_\infty \leq \frac{4}{\pi} \sum_{k=n}^{\infty} \frac{1}{4k^2 - 1} \leq c/n.$$

Since f is not differentiable at $\pm\pi/2$, the converse of the Favard estimate is not true.

The correct device is a regularization functional that is known in this context as a K-functional.

Definition 2.2 If $r \in \mathbb{N}$, $1 \leq p \leq \infty$, and $f \in X^p$, then the K-functional is defined by

$$K_{r,p}(f, \delta) \overset{\text{def}}{=} \inf\{\|f - g\|_p + \delta^r \|g^{(r)}\|_p\}, \qquad (2.20)$$

where the infimum is taken over all g for which $g^{(r)} \in X^p$. If $\gamma > 0$ and $r > \gamma$ is an integer, then we define

$$\|f\|_{\gamma,p} \overset{\text{def}}{=} \sup_{0<\delta<1/2} \frac{K_{r,p}(f, \delta)}{\delta^\gamma}. \qquad (2.21)$$

The set of all functions $f \in X^p$ for which $\|f\|_{\gamma,p} < \infty$ is denoted by $W_{\gamma,p}$.

The following theorem [12, Chapter 7, Theorem 9.2], [57, Chapter III, Section 3.2, Corollary 7] shows the close connection between the apparently somewhat artificially defined class $W_{\gamma,p}$, the rate of decrease of the degrees of approximation, and the smoothness in the classical sense.

Theorem 2.3 *Let $1 \le p \le \infty$, $f \in X^p$, $\gamma > 0$, $r > \gamma$ be an integer, $\gamma = s + \beta$ where $s \ge 0$ is an integer chosen so that $0 < \beta \le 1$, and let h be a twice continuously differentiable low pass filter.*

(a) *$f \in W_{\gamma,p}$ if and only if $E_{n,p}(f) = \mathcal{O}(n^{-\gamma})$. More precisely,*

$$\|f\|_{\gamma,p} \sim \sup_{n \in \mathbb{N}} n^{\gamma} E_{n,p}(f) \sim \sup_{n \in \mathbb{N}} n^{\gamma} \|f - \sigma_n(h, f)\|_p, \qquad (2.22)$$

where the constants involved in "$\sim$" are independent of f.

(b) *We have $f \in W_{\gamma,p}$ if and only if $f^{(s)} \in W_{\beta,p}$.*

We remark that since the quantity $E_{n,p}(f)$ does not depend on the choice of r in the definition of $W_{\gamma,p}$ except that $r > \gamma$, the class $W_{\gamma,p}$ does not depend on the choice of r either, as long as $r > \gamma$. Second, we note that in the above theorem $s \neq \lfloor \gamma \rfloor$. In particular, for the function $x \mapsto |\cos x|$, one has $\gamma = 1$, $s = 0$ and $\beta = 1$. There are characterizations of the K-functional that are given directly in terms of the function f rather than as a regularization functional. For instance (cf. [57, Chapter III, Section 1.2, Theorem 6], [12, Chapter 6, Theorem 2.4]),

$$K_{r,p}(f, \delta) \sim \sup_{|t| \le \delta} \left\| \sum_{k=0}^{r} (-1)^k \binom{r}{k} f(\circ + kt) \right\|_p,$$

where the right hand expression is called the r-th order modulus of smoothness of f, and the constants involved in "$\sim$" are independent of f and δ. In particular, if $0 < \beta < 1$, then $f \in W_{\beta,\infty}$ is equivalent to the Lipschitz–Hölder condition on f ([57, Chapter III, Section 3.2, Corollary 7] [12, Chapter 7, Theorem 3.3]):

$$|f(x + t) - f(x)| \le c\|f\|_{\beta,\infty}|t|^{\beta}, \quad x \in [-\pi, \pi]. \qquad (2.23)$$

In modern approximation theory, it has become more customary to take the K-functional itself as a measurement of smoothness in various situations, taking such direct relationships for granted.

We end this subsection by pointing out an interesting property of the summability operator, referred to in approximation theory literature as realization of the K-functional.

Theorem 2.4 *Let $r \in \mathbb{N}$, $1 \le p \le \infty$, and $f \in X^p$. Let h be a twice continuously differentiable low pass filter. Then for $n \in \mathbb{N}$,*

$$\|f - \sigma_n(h, f)\|_p + n^{-r}\|\sigma_n^{(r)}(h, f)\|_p \sim K_{r,p}(f, 1/n), \qquad (2.24)$$

where the constants involved in "$\sim$" are independent of both f and n.

Thus, when one is not interested in finding the function g that achieves the infimum in the definition of the K-functional, $\sigma_n(h, f)$ supplies a near-optimal solution. Moreover, while finding the minimizer g is a separate optimization problem for each r and p, the summability operator works for *every* r and *every* p, and its construction does not involve any optimization.

The essential ideas behind the proof of Theorem 2.4 are in the paper [8] of Czipszer and Freud. For a lack of easy reference, we include the simple proof.

Proof of Theorem 2.4 The definition (2.20) of the K-functional shows that

$$K_{r,p}(f, 1/n) \leq \|f - \sigma_n(h, f)\|_p + n^{-r}\|\sigma_n^{(r)}(h, f)\|_p.$$

We prove the inequality in the other direction. Let $g \in X^p$ be any function with $g^{(r)} \in X^p$. Then using Theorem 2.1 and the Favard estimate (2.10), we obtain

$$\|f - \sigma_n(h, f)\|_p \leq \|f - g - \sigma_n(h, f - g)\|_p + \|g - \sigma_n(h, g)\|_p \leq \|f - g\|_p$$
$$+\|\sigma_n(h, f - g)\|_p + cE_{n/2}(g) \leq c\left\{\|f - g\|_p + \frac{1}{n^r}\|g^{(r)}\|_p\right\}. \tag{2.25}$$

Further, using the fact that $\sigma_n(h, g)^{(r)} = \sigma_n(h, g^{(r)})$ and the Bernstein inequality (2.9), we deduce that

$$n^{-r}\|\sigma_n^{(r)}(h, f)\|_p \leq n^{-r}\|\sigma_n(h, f - g)^{(r)}\|_p + n^{-r}\|\sigma_n(h, g)^{(r)}\|_p$$
$$\leq c\{\|\sigma_n(h, f - g)\|_p + n^{-r}\|\sigma_n(g^{(r)})\|_p\} \leq c\left\{\|f - g\|_p + \frac{1}{n^r}\|g^{(r)}\|_p\right\}.$$

Together with (2.25), we have shown that

$$\|f - \sigma_n(h, f)\|_p + n^{-r}\|\sigma_n^{(r)}(h, f)\|_p \leq c\left\{\|f - g\|_p + \frac{1}{n^r}\|g^{(r)}\|_p\right\}.$$

Since g is an arbitrary function with $g^{(r)} \in X^p$, the definition (2.20) shows that

$$\|f - \sigma_n(h, f)\|_p + n^{-r}\|\sigma_n^{(r)}(h, f)\|_p \leq cK_{r,p}(f, 1/n).$$

$\square$

2.3 Wavelet-like representation

In this section, h will denote a fixed, infinitely differentiable low pass filter. We have seen that $f \in L^1$ if and only if $\sigma_n(h, f) \to f$ in L^1. Since $\sigma_n(h, f)$ is defined entirely in terms of the sequence $\{\hat{f}(k)\}_{k \in \mathbb{Z}}$, it follows that the sequence of Fourier coefficients of an integrable function determines the function uniquely. For this reason, this sequence is called the *frequency domain description* of f, while a formula giving

f directly as a function of its argument is called a *space (or time) domain description*. There are some problems with the frequency domain description.

Except in the case of the L^2 norm, where the Parseval identity is available, the frequency domain description of a function does not reveal its smoothness directly. For example, we consider

$$f_1(x) \stackrel{\text{def}}{=} \frac{\cos((\pi - x)/4)}{\sqrt{2}\sin(x/2)},$$

for which the formal Fourier series expansion is given by

$$1 + \frac{2}{\sqrt{\pi}} \sum_{k=1}^{\infty} \frac{\Gamma(k + 1/2)}{k!} \cos(kx),$$

see, e.g., [79, Chapter V, Formula (2.3)], so that, using Stirling's formula, $\hat{f}_1(k) = \mathcal{O}(k^{-1/2})$. The function f_1 is discontinuous at 0. On the other hand, for

$$f_2(x) \stackrel{\text{def}}{=} \frac{1}{2} + \sum_{j=1}^{\infty} \frac{\cos(4^j x)}{2^j},$$

it is easy to verify using Theorem 2.3 that $f_2 \in W_{1/2,\infty}$ even though $\hat{f}_2(k) = \mathcal{O}(k^{-1/2})$ as well. For the function

$$f_3(x) \stackrel{\text{def}}{=} |\cos x|^{1/2} = \frac{\Gamma(3/2)}{\sqrt{2}\Gamma(5/4)^2} + \sum_{j=1}^{\infty} (-1)^j \frac{\sqrt{2}\Gamma(3/2)}{\Gamma(-1/4)\Gamma(5/4)} \frac{\Gamma(j-1/4)}{\Gamma(j+3/4)} \cos(2jx),$$

see, e.g., [17, p. 12, Eqn. (30)], the formulation (2.23) of the class $W_{1/2,\infty}$ shows that $f_3 \in W_{1/2,\infty}$, but $f_3 \notin W_{\gamma,\infty}$ for any $\gamma > 1/2$. However, $\hat{f}_3(k) \sim k^{-3/2}$. Finally, f_2 is nowhere differentiable, while f_3 and f_1 both admit analytic continuations at all but finitely many points on $\mathbb{T}$.

In the last couple of decades, wavelet analysis has become popular as an alternative to Fourier series where the coefficients characterize local smoothness of the target function, see [10, Theorems 9.2.1, 9.2.2]. We find it interesting both from a theoretical as well as practical point of view to develop a similar expansion which can be computed using the classical Fourier coefficients, but achieves the same purpose. In this section, we review some of our results in this direction, sketching some proofs related to the summability kernel. This section is based on [58–60].

We pause again to discuss an example from [34] for the utility of the summability kernel and its localization. In Fig. 3 below, we report the log-plot of the error between $x \mapsto |\cos x|^{1/4}$ and its Fourier projection of degree 31, compared to the error obtained by using our summability operator with a smooth filter h. In keeping with the converse theorem, the maximum error is of the same order of magnitude in both the cases, but it is clear that the error using the summability operator decreases rapidly as $|x - \pi/2|$

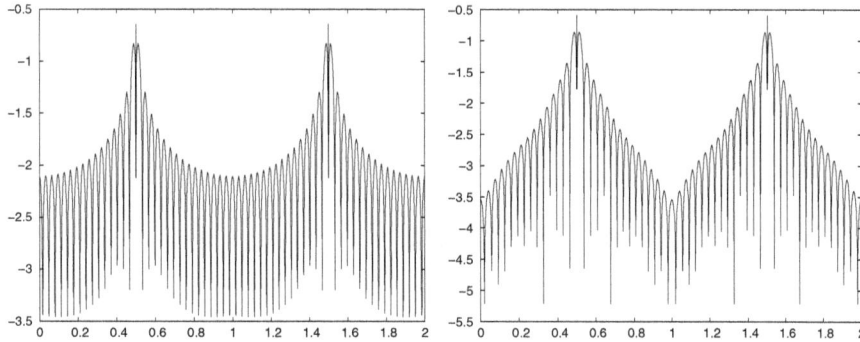

Fig. 3 The plot of the logarithm (base 10) of the absolute error between the function $x \mapsto |\cos x|^{1/4}$, and (*left*) its Fourier projection (*right*) trigonometric polynomial obtained by our summability operator, where the Fourier coefficients are estimated by 128 point FFT. The order of the trigonometric polynomials is 31 in each case. The numbers on the x axis are in multiples of π, the actual absolute errors are 10^y

or $|x + \pi/2|$ increases. Thus, the summability operator is more robust with respect to local "singularities" in a function.

To resume our main discussion, we note that unlike the notion of derivatives, membership in $W_{\gamma,p}$ is not a local property. The following is a standard way to define such spaces locally. We refer to an open connected subset of $\mathbb{T}$ as an *arc*. Essentially, an arc is an open subinterval of $[-\pi, \pi]$, except that an arc might also be of the form $[-\pi, a) \cup (b, \pi]$. The class of all infinitely differentiable functions supported on an arc I will be denoted by C_I^∞.

Definition 2.3 Let $1 \le p \le \infty$, $\gamma > 0$, $x_0 \in \mathbb{T}$, and $f \in L^1$. We say that $f \in W_{\gamma,p}(x_0)$ if there exists an arc $I \ni x_0$ such that for every $\phi \in C_I^\infty$, $\phi f \in W_{\gamma,p}$.

Thus, among the three functions listed above, $f_1 \in W_{\gamma,\infty}(x_0)$ for all $\gamma > 0$ and $x_0 \ne 0$, $f_2 \notin W_{\gamma,\infty}(x_0)$ for any x_0 if $\gamma > 1/2$, and $f_3 \in W_{\gamma,\infty}(x_0)$ for all $\gamma > 0$, $x_0 \ne \pm\pi/2$.

We define the frame operators

$$\tau_j(h, f) \overset{\text{def}}{=} \begin{cases} \sigma_1(h, f), & \text{if } j = 0, \\ \sigma_{2^j}(h, f) - \sigma_{2^{j-1}}(h, f), & \text{if } j \in \mathbb{N}. \end{cases} \quad (2.26)$$

Note that these are the same as the polynomials P_j in (2.11). We let

$$\Psi_j(h, t) \overset{\text{def}}{=} \Phi_{2^{j+1}}(h, t) - \Phi_{2^{j-2}}(h, t), \quad j = 0, 1, \ldots, \quad (2.27)$$

where we recall our convention that $\Phi_n(h, t) = 0$ if $n < 1$.

We will not use the following constructions in the rest of the paper, but state them for the sake of completeness. Let

$$g^*(t) \overset{\text{def}}{=} \sqrt{h(t) - h(2t)}. \quad (2.28)$$

We write for $t \in \mathbb{T}$, and $f \in L^1$,

$$
\Psi_j^*(t) \overset{\text{def}}{=} \begin{cases} \Phi_{2^j}(g^*, t), & j \in \mathbb{N}, \\ 1, & j = 0, \end{cases} \tag{2.29}
$$

and

$$
\tau_j^*(h, f) \overset{\text{def}}{=} \sigma_{2^j}(g^*, f), \quad j = 1, 2, \ldots, \quad \tau_0^*(h, f, t) \overset{\text{def}}{=} \hat{f}(0). \tag{2.30}
$$

Finally, we set

$$
v_{k,j} = k\pi/2^j, \quad k = -2^j + 1, \ldots, 2^j, \quad j = 0, 1, 2, \ldots. \tag{2.31}
$$

It can be verified using the formula for the sum of geometric series that

$$
\frac{1}{2^{j+1}} \sum_{k=-2^j+1}^{2^j} \exp(i\ell v_{k,j}) = \begin{cases} 0, & \text{if } \ell = \pm 1, \ldots, \pm(2^{j+1} - 1), \\ 1, & \text{if } \ell = 0, \end{cases}
$$

and hence, the following quadrature formula holds.

$$
\frac{1}{2\pi} \int_{\mathbb{T}} P(t)\, dt = \frac{1}{2^{j+1}} \sum_{k=-2^j+1}^{2^j} P(v_{k,j}), \quad P \in \mathbb{H}_{2^{j+1}}. \tag{2.32}
$$

The following theorem states the wavelet-like (frame) expansion of functions in X^p. We note that, unlike classical Littlewood–Paley expansions or wavelet expansions, our results are valid also for $f \in X^p$, where $p = 1$ and $p = \infty$ are included. We note that the coefficients $\tau_j(h, f, v_{k,j})$ and $\tau_j^*(h, f, v_{k,j})$ can be computed as finite linear combinations of the Fourier coefficients of f.

Theorem 2.5 *Let $1 \leq p \leq \infty$, $\gamma > 0$, $f \in X^p$, and let h be an infinitely differentiable low pass filter.*

(a) *We have*

$$
f = \sum_{j=0}^{\infty} \tau_j(h, f) = \sum_{j=0}^{\infty} \frac{1}{2^{j+1}} \sum_{k=-2^j+1}^{2^j} \tau_j(h, f, v_{k,j}) \Psi_j(h, \circ - v_{k,j})
$$

$$
= \sum_{j=0}^{\infty} \frac{1}{2^{j+1}} \sum_{k=-2^j+1}^{2^j} \tau_j^*(h, f, v_{k,j}) \Psi_j^*(h, \circ - v_{k,j}), \tag{2.33}
$$

with convergence in the sense of X^p.

(b) *If $f \in L^2$ then*

$$\|f\|_2^2 \leq 2 \sum_{j=0}^{\infty} \|\tau_j(h, f)\|_2^2 = \sum_{j=0}^{\infty} \frac{1}{2^j} \sum_{k=-2^j+1}^{2^j} |\tau_j(h, f, v_{k,j})|^2 \leq 2\|f\|_2^2 \quad (2.34)$$

and

$$\|f\|_2^2 = \sum_{j=0}^{\infty} \|\tau_j^*(h, f)\|_2^2 = \sum_{j=0}^{\infty} \frac{1}{2^{j+1}} \sum_{k=-2^j+1}^{2^j} |\tau_j^*(h, f, v_{k,j})|^2. \quad (2.35)$$

Proof The first equation in (2.33) follows from (2.8) by a telescoping series argument, cf. the proof of (2.13) (the polynomials denoted in the latter by P_j are in fact $\tau_j(h, f)$). A comparison of Fourier coefficients shows that for $j = 0, 1 \ldots,$

$$\tau_j(h, f, x) = \frac{1}{2\pi} \int_{\mathbb{T}} \tau_j(h, f, t) \Psi_j(h, x - t) \, dt$$

$$= \frac{1}{2\pi} \int_{\mathbb{T}} \tau_j^*(h, f, t) \Psi_j^*(h, x - t) \, dt.$$

The quadrature formula (2.32) can be used to express $\tau_j(h, f)$ in the first equation in (2.33) as the inner sums as indicated in the remaining two equations there.

We observe that $h(t) - h(2t) \geq 0$ for all t and $h(t) - h(2t) \neq 0$ only if $1/4 < t < 1$. So, for any $k \in \mathbb{Z}\backslash\{0\}$, if m_k is the integer part of $\log_2 |k|$, then

$$h\left(\frac{k}{2^j}\right) - h\left(\frac{k}{2^{j-1}}\right) \neq 0 \text{ only if } j = m_k + 1 \text{ or } j = m_k + 2$$

Consequently, recalling that $0 \leq h(t) \leq 1$ for all $t \in \mathbb{R}$, we obtain for $k \in \mathbb{Z}\backslash\{0\}$ that

$$1 = \left(\sum_{j=1}^{\infty} h\left(\frac{k}{2^j}\right) - h\left(\frac{k}{2^{j-1}}\right)\right)^2 \leq 2\sum_{j=1}^{\infty}\left(h\left(\frac{k}{2^j}\right) - h\left(\frac{k}{2^{j-1}}\right)\right)^2$$

$$\leq 2\sum_{j=1}^{\infty} h\left(\frac{k}{2^j}\right) - h\left(\frac{k}{2^{j-1}}\right) = 2. \quad (2.36)$$

Using the Parseval identity and Fubini's theorem, we see that

$$\sum_{j=0}^{\infty} \|\tau_j(h, f)\|_2^2 = |\hat{f}(0)|^2 + \sum_{j=1}^{\infty} \sum_{k \in \mathbb{Z}\backslash\{0\}} \left(h\left(\frac{k}{2^j}\right) - h\left(\frac{k}{2^{j-1}}\right)\right)^2 |\hat{f}(k)|^2$$

$$= |\hat{f}(0)|^2 + \sum_{k \in \mathbb{Z}\backslash\{0\}} |\hat{f}(k)|^2 \sum_{j=1}^{\infty} \left(h\left(\frac{k}{2^j}\right) - h\left(\frac{k}{2^{j-1}}\right)\right)^2.$$

Using the Parseval identity again and (2.36), we conclude that

$$\|f\|_2^2 = \sum_{k \in \mathbb{Z}} |\hat{f}(k)|^2 \le 2 \sum_{j=0}^{\infty} \|\tau_j(h, f)\|_2^2 \le 2\|f\|_2^2. \qquad (2.37)$$

The quadrature formula (2.32) shows that

$$\|\tau_j(h, f)\|_2^2 = \frac{1}{2^{j+1}} \sum_{k=-2^j+1}^{2^j} |\tau_j(h, f, v_{k,j})|^2, \quad j = 0, 1, \ldots.$$

Together with (2.37), this completes the proof of (2.34).

Another simple application of the Parseval identity gives

$$\|f\|_2^2 = \sum_{j=0}^{\infty} \|\tau_j^*(h, f)\|_2^2,$$

which, together with the quadrature formula (2.32) leads to (2.35). $\square$

The following theorem shows that the coefficients of the expansions in (2.33) provide a complete characterization of local smoothness classes, analogous to the corresponding theorems in wavelet analysis.

Theorem 2.6 *Let* $1 \le p \le \infty$, $\gamma > 0$, $x_0 \in \mathbb{T}$, $f \in X^p$, *and let* h *be an infinitely differentiable low pass filter. Then the following statements are equivalent.*

(a) $f \in W_{\gamma, p}(x_0)$.

(b) *There exists an arc* $I \ni x_0$ *such that*

$$\left\{ \frac{1}{2^{j+1}} \sum_{k:v_{k,j} \in I} |\tau_j(h, f, v_{k,j})|^p \right\}^{1/p} = \mathcal{O}(2^{-j\gamma}), \quad 1 \le p < \infty. \quad (2.38)$$

In the case when $p = \infty$, *the above estimate is interpreted as*

$$\max_{k:v_{k,j} \in I} |\tau_j(h, f, v_{k,j})| = \mathcal{O}(2^{-j\gamma}). \qquad (2.39)$$

(c) *There exists an arc* I *containing* x_0 *such that*

$$\left\{ \frac{1}{2^{j+1}} \sum_{k:v_{k,j} \in I} |\tau_j^*(h, f, v_{k,j})|^p \right\}^{1/p} = \mathcal{O}(2^{-j\gamma}), \qquad (2.40)$$

with a modification as done in (b) in the case when $p = \infty$.

We will prove only the equivalence of parts (a) and (b) above; the equivalence between parts (b) and (c) do not add much to the concepts we wish to emphasize here. This proof depends on the following Marcinkiewicz–Zygmund–type inequalities, see [79, Chapter X, Theorem (7.5) and the remark thereafter, and Theorem (7.28)].

Lemma 2.1 *For $1 \le p \le \infty$ and $T \in \mathbb{H}_{2^j+1}$, we have*

$$\|T\|_p \sim \left\{ \frac{1}{2^{j+1}} \sum_{k=-2^j+1}^{2^j} |T(v_{k,j})|^p \right\}^{1/p}, \tag{2.41}$$

where the usual interpretation for the middle expression is assumed when $p = \infty$, and the constants involved in "$\sim$" are independent of j and T.

Proof of (a)$\Longleftrightarrow$(b) *of Theorem 2.6* In this proof, we will write for any arc I, integer $j \ge 1$ and $g : \mathbb{T} \to \mathbb{R}$,

$$[g]_{j,p,I} = \left\{ \frac{1}{2^{j+1}} \sum_{k:v_{k,j} \in I} |g(v_{k,j})|^p \right\}^{1/p}, \quad 1 \le p \le \infty.$$

Let $f \in W_{\gamma,p}(x_0)$, and $J \ni x_0$ be an arc such that $\phi f \in W_{p,\gamma}$ for every infinitely differentiable function ϕ supported on J. We consider subarcs $I \subset I' \subset J$ with $x_0 \in I$, and an infinitely differentiable function ψ which is equal to 1 on I' and 0 outside J. We choose and fix an integer $S > \max(1,\gamma)$. Since h is infinitely differentiable, it follows from (2.6) that

$$|\Psi_j(t)| \le c2^j \min\left(1, (2^j|t|)^{-S-1}\right), \quad t \in \mathbb{T}, \ j \ge 0. \tag{2.42}$$

Therefore, for $x \in I$,

$$|\tau_j((1-\psi)f,x)| \le \frac{1}{2\pi} \int_{\mathbb{T}} |(1-\psi(t))f(t)\Psi_j(x-t)|\,dt$$
$$= \frac{1}{2\pi} \int_{\mathbb{T}\setminus I'} |(1-\psi(t))f(t)\Psi_j(x-t)|\,dt \le c(I,I')2^{-jS}\|f\|_1,$$

and

$$|\tau_j(f,x)| \le |\tau_j(\psi f,x)| + |\tau_j((1-\psi)f,x)| \le |\tau_j(\psi f,x)| + c(I,I')2^{-jS}\|f\|_1.$$

Consequently, using Lemma 2.1, we conclude that

$$[\tau_j(f)]_{j,p,I} \le [\tau_j(\psi f)]_{j,p,I} + c(I,I')2^{-jS}\|f\|_1$$
$$\le [\tau_j(\psi f)]_{j,p,\mathbb{T}} + c(I,I')2^{-jS}\|f\|_1$$
$$\le c(I,I')\{\|\tau_j(\psi f)\|_p + 2^{-jS}\|f\|_1\}. \tag{2.43}$$

Since $\tau_j(P) \equiv 0$ for all $P \in \mathbb{H}_{2^{j-1}}$, the boundedness of the operators τ_j implies that for any $P \in \mathbb{H}_{2^{j-1}}$,

$$\|\tau_j(\psi f)\|_p = \|\tau_j(\psi f - P)\|_p \le c\|\psi f - P\|_p,$$

and, hence,

$$\|\tau_j(\psi f)\|_p \le cE_{2^{j-1},p}(\psi f).$$

Therefore, (2.43) shows that

$$[\tau_j(f)]_{j,p,I} \le c(I, I')\{E_{2^{j-1},p}(\psi f) + 2^{-jS}\|f\|_1\}.$$

Since $\psi f \in W_{\gamma,p}$ and $S > \gamma$, an application of Theorem 2.3 now leads to part (b).

Conversely, let $I \ni x_0$ be an arc such that

$$[\tau_j(f)]_{j,p,I} = \mathcal{O}(2^{-j\gamma}), \tag{2.44}$$

and let ϕ be any infinitely differentiable function supported on I. The Favard estimate then shows that for every $j \ge 1$ there exists $R_j \in \mathbb{H}_{2^j}$ such that

$$\|\phi - R_j\|_\infty \le c(\phi)2^{-jS}. \tag{2.45}$$

Therefore, in view of (2.7), we obtain

$$E_{2^{j+1},p}(\phi f) \le \|\phi f - R_j \sigma_{2^j}(h, f)\|_p \le \|\phi(f - \sigma_{2^j}(h, f))\|_p + \|(\phi - R_j)\sigma_{2^j}(h, f)\|_p$$

$$\le \|\phi(f - \sigma_{2^j}(h, f))\|_p + c(\phi)2^{-jS}\|f\|_p. \tag{2.46}$$

Using (2.33), Lemma 2.1, (2.45), and (2.7) in that order, we see that

$$\|\phi(f - \sigma_{2^j}(h, f))\|_p = \left\| \phi \sum_{k=j+1}^{\infty} \tau_k(h, f) \right\|_p \le \sum_{k=j+1}^{\infty} \|\phi\tau_k(h, f)\|_p$$

$$\le \sum_{k=j+1}^{\infty} \|R_k\tau_k(h, f)\|_p + \sum_{k=j+1}^{\infty} \|(\phi - R_k)\tau_k(h, f)\|_p$$

$$\le c\left\{ \sum_{k=j+1}^{\infty} [R_k\tau_k(h, f)]_{k,p,\mathbb{T}} + \sum_{k=j+1}^{\infty} \|(\phi - R_k)\tau_k(h, f)\|_p \right\}$$

$$\le c\left\{ \sum_{k=j+1}^{\infty} [\phi\tau_k(h, f)]_{k,p,\mathbb{T}} + \sum_{k=j+1}^{\infty} [(\phi - R_k)\tau_k(h, f)]_{k,p,\mathbb{T}} + \sum_{k=j+1}^{\infty} \|(\phi - R_k)\tau_k(h, f)\|_p \right\}$$

$$\le c(\phi)\left\{ \sum_{k=j+1}^{\infty} [\tau_k(h, f)]_{k,p,I} + \sum_{k=j+1}^{\infty} 2^{-kS}[\tau_k(h, f)]_{k,p,\mathbb{T}} + \sum_{k=j+1}^{\infty} 2^{-kS}\|\tau_k(h, f)\|_p \right\}$$

$$\leq c(\phi) \left\{ \sum_{k=j+1}^{\infty} [\tau_k(h, f)]_{k,p,I} + \sum_{k=j+1}^{\infty} 2^{-kS} \|\tau_k(h, f)\|_p \right\}$$

$$\leq c(\phi) \left\{ \sum_{k=j+1}^{\infty} [\tau_k(h, f)]_{k,p,I} + \sum_{k=j+1}^{\infty} 2^{-kS} \|f\|_p \right\}.$$

Thus, the assumption (2.44) leads to

$$\|\phi(f - \sigma_{2^j}(h, f))\|_p = \mathcal{O}(2^{-j\gamma}).$$

Since $S > \gamma$, this estimate and (2.46) show that $E_{2^{j+1},p}(\phi f) = \mathcal{O}(2^{-j\gamma})$. Therefore, Theorem 2.3 implies that for every infinitely differentiable ϕ supported on I we have $\phi f \in W_{\gamma,p}$, that is, (a) holds.   □

We note that an expansion of the form (2.33) is used often in approximation theory. We reserve the term *wavelet-like representation* to indicate that the behavior of the terms of the expansion characterize **local smoothness** of the target function. Thus, for example, although the expansion

$$f = \sum_{j=0}^{\infty} (v_{2^{j+1}}(f) - v_{2^j}(f)) + v_1(f)$$

is very similar to (2.33), and holds for every $f \in X^p$ $p \in [1, \infty]$, we do not refer to this expansion as a wavelet-like representation, because the localization properties of the operators v_n are not strong enough to admit an analogue of Theorem 2.6 for characterization of local smoothness classes for the smoothness parameter >1.

3 Multivariate analogues

3.1 Notation

In what follows, $q \geq 2$ is a fixed integer and the various constants will depend on q. As usual, the notation $\mathbb{T}^q$ denotes the q dimensional torus, that is, a q-fold cross product of $\mathbb{T}$ with itself. As in the univariate case, functions on $\mathbb{T}^q$ can be considered as functions on $\mathbb{R}^q$, 2π-periodic in each variable. An *arc* in this context has the form $\prod_{j=1}^{q} I_k$, where each $I_k \subseteq \mathbb{T}$ is a univariate arc as defined in Sect. 2. An element of $\mathbb{R}^q$ will be denoted in bold face, e.g., $\mathbf{x} = (x_1, \ldots, x_q)$. We find it convenient to write x also for the vector $(x, x, \ldots, x)$; e.g., 0 denotes both the scalar 0 and the vector $(0, 0, \ldots, 0)$. We hope that it will be clear from the context whether a scalar is intended or a vector with equal components is intended. When applied to vectors, univariate operations and relations will be interpreted in a coordinatewise sense; e.g., $\mathbf{x} \geq 0$ means that $x_j \geq 0$ for all $j = 1, 2, \ldots, q$, $\mathbf{x}^{\mathbf{y}} = (x_1^{y_1}, \ldots, x_q^{y_q})$, whenever the expressions are defined, and so forth. The notation $\mathbf{x} \cdot \mathbf{y}$ denotes the inner product between $\mathbf{x}$ and $\mathbf{y}$. For $0 < p \le \infty$, we define

$$|\mathbf{x}|_p = \begin{cases} \left(\sum_{k=1}^q |x_k \bmod (2\pi)|^p\right)^{1/p}, & \text{if } 0 < p < \infty, \\ \max_{1 \le k \le q} |x_k \bmod (2\pi)|, & \text{if } p = \infty. \end{cases}$$

In general, many of the definitions of norms and related expressions, are very similar to those in the univariate case, and we will often omit the dimension q whenever we feel that writing it explicitly makes the notation unnecessarily cumbersome.

If $f : \mathbb{T}^q \to \mathbb{R}$ is differentiable, we write $\partial_r f$ to denote the partial derivative of f with respect to the r-th variable. If $\mathbf{r} \in \mathbb{Z}_+^q$ and f is sufficiently smooth, we write $\partial^{\mathbf{r}} f$ to denote the partial derivative indicated by $\mathbf{r}$.

The measure μ_q^* denotes the Lebesgue measure on $\mathbb{T}^q$ normalized to 1. If $f : \mathbb{T}^q \to \mathbb{C}$ is Lebesgue measurable, and $A \subseteq \mathbb{T}^q$ is Lebesgue measurable, we write

$$\|f\|_{p,A} \stackrel{\text{def}}{=} \begin{cases} \left\{\int_A |f(\mathbf{x})|^p d\mu_q^*(\mathbf{x})\right\}^{1/p}, & \text{if } 0 < p < \infty, \\ \mu_q^*\text{-ess sup}_{\mathbf{x} \in A} |f(\mathbf{x})|, & \text{if } p = \infty. \end{cases} \tag{3.1}$$

As before, if $A = \mathbb{T}^q$, then it will be omitted from the notation; e.g., $\|f\|_p = \|f\|_{p,\mathbb{T}^q}$. The spaces L^p are defined as usual.

3.2 Approximation theory

There are many ways to define trigonometric polynomials of several variables. In particular, for p such that $0 < p \le \infty$, a trigonometric polynomial of ℓ^p-degree $< n$ is a function of the form

$$\mathbf{x} \mapsto \sum_{\mathbf{k}: |\mathbf{k}|_p < n} a_{\mathbf{k}} \exp(i\mathbf{k} \cdot \mathbf{x}).$$

When $p = \infty$, we refer to *coordinatewise-degree* $<n$, whereas if $p = 1$, we refer to *total degree* $<n$, and when $p = 2$, we refer to the *spherical degree* $<n$. The dimensions of all these spaces are always $\mathcal{O}(n^q)$. In our discussion below, we will restrict ourselves to the class $\mathbb{H}_n^q$ of trigonometric polynomials of spherical degree $<n$, but the results will be equally valid also for other values of p. The L^p closure of $\cup_{n \ge 0} \mathbb{H}_n^q$ will be denoted by X^p (or $X^p(\mathbb{T}^q)$ if some confusion is likely to result).

If $f \in X^p(\mathbb{T}^q)$ and $n \ge 0$, then the degree of approximation of f from $\mathbb{H}_n^q$ is defined by

$$E_{n,p}(f) \stackrel{\text{def}}{=} E_{q;n,p}(f) \stackrel{\text{def}}{=} \inf\{\|f - P\|_p : P \in \mathbb{H}_n^q\}.$$

Again, the symbol q will be omitted when we don't expect any confusion. If $f \in L^1$, its Fourier coefficient is defined by

$$\hat{f}(\mathbf{k}) = \int\limits_{\mathbb{T}^q} f(\mathbf{t}) \exp(i\mathbf{k} \cdot \mathbf{t}) \, d\mu_q^*(\mathbf{t}), \quad \mathbf{k} \in \mathbb{Z}^q.$$

Let $H : \mathbb{R}^q \to \mathbb{R}$ be compactly supported. The role of the kernel Φ_n is played by

$$\Phi_n(H, \mathbf{t}) \overset{\text{def}}{=} \Phi_{n,q}(H, \mathbf{t}) \overset{\text{def}}{=} \sum_{\mathbf{k} \in \mathbb{Z}^q} H\left(\frac{\mathbf{k}}{n}\right) \exp(i\mathbf{k} \cdot \mathbf{t}), \quad n \in \mathbb{N}, \ \mathbf{t} \in \mathbb{T}^q,$$

and the corresponding operator is defined by

$$\sigma_n(H, f, \mathbf{x}) = \int\limits_{\mathbb{T}^q} f(\mathbf{t})\Phi_n(H, \mathbf{x} - \mathbf{t}) \, d\mu_q^*(\mathbf{t})$$

$$= \sum_{\mathbf{k} \in \mathbb{Z}^q} H\left(\frac{\mathbf{k}}{n}\right) \hat{f}(\mathbf{k}) \exp(i\mathbf{k} \cdot \mathbf{x}), \quad n \in \mathbb{N}, \ \mathbf{x} \in \mathbb{T}^q.$$

Of particular interest is the case when $H(\mathbf{t}) = h(|\mathbf{t}|_2)$ for some compactly supported $h : [0, \infty) \to \mathbb{R}$. We will overload the notation again, and denote the corresponding kernel and operator by $\Phi_n(h, \circ)$ and $f \mapsto \sigma_n(h, f)$, with the domain of Φ_n and f making it clear which meaning is intended. We observe that when the mapping $\mathbf{x} \mapsto h(|\mathbf{x}|_2)$ is integrable on $\mathbb{R}^q$, then its Fourier transform is also *radial*, that is, it has the form $\mathbf{x} \mapsto F(|\mathbf{x}|_2)$ for some $F : [0, \infty) \to \mathbb{C}$ [75, Chapter IV, Theorem 3.3]. Therefore, if the Poisson summation formula holds, then $\Phi_n(h, \circ)$ is a radial function, 2π-periodic in each of its variables.

The analogue of Theorem 2.1 is the following, where the conditions on the filter H are a bit stronger to ensure the validity of the Poisson summation formula enlisted in the proof of the theorem.

Theorem 3.1 *Let $S > q$ be an integer and let $H : \mathbb{R}^q \to \mathbb{R}$ be an S-times continuously differentiable, compactly supported function.*

(a) *We have*

$$|\Phi_n(H, \mathbf{t})| \le c(H)n^q \min\left(1, (n|\mathbf{t}|_2)^{-S}\right), \quad \mathbf{t} \in \mathbb{T}, \ n \ge 1. \tag{3.2}$$

(b) *For $1 \le p \le \infty$,*

$$\|\sigma_n(H, f)\|_p \le c(H)\|f\|_p, \quad f \in L^p, \ n \in \mathbb{N}. \tag{3.3}$$

(c) *Let $h : \mathbb{R} \to [0, 1]$ be an S times continuously differentiable low pass filter and $n \in \mathbb{N}$. Then $\sigma_n(h, P) = P$ for all $P \in \mathbb{H}_{n/2}^q$. Moreover, for $1 \le p \le \infty$ and $f \in L^p$,*

$$E_{n,p}(f) \leq \|f - \sigma_n(h, f)\|_p \leq c(h)E_{n/2,p}(f). \tag{3.4}$$

Proof The proof of parts (a) and (b) are given in [5, Section 6.1]. To prove part (c), we let $H_1(\mathbf{x}) = h(|\mathbf{x}|_2)$ and observe that since h is constant in a neighborhood of 0,

$$\nabla H_1(\mathbf{x}) = \frac{\mathbf{x}}{|\mathbf{x}|_2}h'(|\mathbf{x}|_2) = 0 \quad \text{in a neighborhood of 0.}$$

Since h is S-times continuously differentiable, it follows that H_1 is as well. Since $\sigma_n(h, f) = \sigma_n(H_1, f)$, the estimate (3.3) holds with H_1 in place of H, that is, $\sigma_n(h, f)$ in place of $\sigma_n(H, f)$. The remainder of the proof is verbatim the same as the corresponding part of Theorem 2.1.   □

We observe an important corollary of Theorem 3.1.

Corollary 3.1 *If $n \in \mathbb{N}$ and $T \in \mathbb{H}_n^q$, then for $1 \leq p \leq \infty$ and $1 \leq r \leq q$,*

$$\|\partial_r T\|_p \leq cn\|T\|_p. \tag{3.5}$$

Proof Since $\widehat{\partial_r T}(\mathbf{k}) = ik_r\hat{T}(\mathbf{k})$ for $\mathbf{k} \in \mathbb{Z}^q$, we may use Theorem 3.1 with an appropriate smooth low pass filter h, and the function $H(\mathbf{x}) = x_r h(|\mathbf{x}|_2)$ as in the proof of the univariate Bernstein inequality, see Theorem 2.2(a), to prove (3.5).

We will denote the Laplacian operator by $\Delta \overset{\text{def}}{=} \sum_{j=1}^q \partial_j^2$, and observe that for a sufficiently smooth f,

$$\widehat{(-\Delta f)}(\mathbf{k}) = |\mathbf{k}|_2^2 \hat{f}(\mathbf{k}),$$

for each $\mathbf{k} \in \mathbb{Z}^q$. For $r > 0$, we define the differential operator $(-\Delta)^{r/2}$ formally by

$$\widehat{(-\Delta)^{r/2}f}(\mathbf{k}) \overset{\text{def}}{=} |\mathbf{k}|_2^r \hat{f}(\mathbf{k}) \tag{3.6}$$

for $f \in L^1$ and $\mathbf{k} \in \mathbb{Z}^q$. Clearly, if $T \in \mathbb{H}_n^q$ for some $n \in \mathbb{N}$, then $(-\Delta)^{r/2}T$ is well defined for all $r > 0$.

With respect to these operators, the analogous Favard estimate and Bernstein inequality are the following.

Theorem 3.2 (a) *Let $1 \leq p \leq \infty$ and let r be a positive even integer. Then for all $f \in X^p$ for which $(-\Delta)^{r/2}f \in X^p$, we have*

$$E_{n,p}(f) \leq cn^{-r}\|(-\Delta)^{r/2}f\|_p.$$

(b) *Let $T \in \mathbb{H}_n^q$ and let r be a positive even integer. Then for $1 \leq p \leq \infty$,*

$$\|(-\Delta)^{r/2}T\|_p \leq cn^r\|T\|_p.$$

The K-functional appropriate for the multivariate theory is defined (with another overload of notation) as follows.

Definition 3.1 If $r \geq 1$ is an even integer, $1 \leq p \leq \infty$, and $f \in X^p$, then we define

$$K_{r,p}(f,\delta) \overset{\text{def}}{=} K_{q;r,p}(f,\delta) \overset{\text{def}}{=} \inf\{\|f - g\|_p + \delta^r \|(-\Delta)^{r/2}g\|_p\}, \qquad (3.7)$$

where the infimum is taken over all g for which $(-\Delta)^{r/2}g \in X^p$. If $\gamma > 0$ and $r > \gamma$ is an even integer, then we define

$$\|f\|_{\gamma,p} \overset{\text{def}}{=} \|f\|_{q;\gamma,p} \overset{\text{def}}{=} \sup_{0 < \delta < 1/2} \frac{K_{r,p}(f,\delta)}{\delta^\gamma}. \qquad (3.8)$$

The set of all functions $f \in X^p$ for which $\|f\|_{\gamma,p} < \infty$ is denoted by $W_{\gamma,p}$ or, if some confusion is likely to happen, then by $W_{q;\gamma,p}$.

The following theorem is the direct analogue of Theorem 2.3 and it is proved in the same way.

Theorem 3.3 *Let* $1 \leq p \leq \infty$, $f \in X^p$, $\gamma > 0$, $r > \gamma$ *be an even integer,* $\gamma = s + \beta$ *where* $s \geq 0$ *is an integer chosen so that* $0 < \beta \leq 1$. *Let* $S > q$ *be an integer and* h *be an* S-*times continuously differentiable low pass filter.*

(a) $f \in W_{\gamma,p}$ *if and only if* $E_{n,p}(f) = \mathcal{O}(n^{-\gamma})$. *More precisely,*

$$\|f\|_{\gamma,p} \sim \sup_{n \in \mathbb{N}} n^\gamma E_{n,p}(f) \sim \sup_{n \in \mathbb{N}} n^\gamma \|f - \sigma_n(h,f)\|_p, \qquad (3.9)$$

where the constants involved in "$\sim$" are independent of f.
(b) *We have* $f \in W_{\gamma,p}$ *if and only if* $(-\Delta)^{s/2}f \in W_{\beta,p}$.

Finally, we state the following theorem.

Theorem 3.4 *Let* $1 \leq p \leq \infty$, $f \in X^p$, *and let* $r \geq 1$ *be an even integer. Let* $S > q$ *be an integer and let* h *be an* S-*times continuously differentiable low pass filter. Then for* $n \in \mathbb{N}$,

$$\|f - \sigma_n(h,f)\|_p + n^{-r}\|(-\Delta)^{r/2}\sigma_n(h,f)\|_p \sim K_{r,p}(f, 1/n), \qquad (3.10)$$

where the constants involved in "$\sim$" are independent of f *and* n.

3.3 Discretization

For many applications in learning theory, the information available about the target function f consists of its values $f(\mathbf{y}_k)$ at finitely many points $\{\mathbf{y}_k\}_{k=1}^M$ but one cannot prescribe in advance the precise location of these points. The goal of this section is to survey the ideas behind a construction of summability operators based on such information which have properties similar to those of the operators $\sigma_n(h)$. An essential ingredient is to obtain real numbers w_k such that for an integer $N \geq 0$ as high as possible, both the *quadrature formula* (3.12) and *M–Z (Marcinkiewicz–Zygmund) inequalities* (3.13) below hold. It will be shown in Theorem 3.5 that such weights can always be found with the desired degree N being dependent on the so-called *density content* of the set $\{\mathbf{y}_k\}$ [43].

Definition 3.2 Let $\mathcal{C} \overset{\text{def}}{=} \{\mathbf{y}_k\}_{k=1}^M \subset \mathbb{T}^q$. We define the *density content* $\delta(\mathcal{C})$ and the *minimal separation* $\eta(\mathcal{C})$ by

$$\delta(\mathcal{C}) = \max_{\mathbf{x} \in \mathbb{T}^q} \min_{1 \leq k \leq M} |\mathbf{x} - \mathbf{y}_k|_\infty, \quad \eta(\mathcal{C}) \overset{\text{def}}{=} \min_{1 \leq k \neq j \leq M} |\mathbf{y}_k - \mathbf{y}_j|_\infty. \quad (3.11)$$

We note that the density content has been referred to in the literature also as *fill distance* or *mesh norm*.

Theorem 3.5 *Let* $\mathcal{C} \overset{\text{def}}{=} \{\mathbf{y}_k\}_{k=1}^M \subset \mathbb{T}^q$ *and* $\delta(\mathcal{C}) \leq 1$. *There exists a positive constant* α *dependent only on* q *with the property that there are non-negative numbers* $\{w_k\}_{k=1}^M$ *such that for* $N \leq \alpha\delta(\mathcal{C})^{-1}$ *we have*

$$\sum_{k=1}^M w_k P(\mathbf{y}_k) = \int_{\mathbb{T}^q} P(\mathbf{t}) d\mu_q^*(\mathbf{t}), \quad P \in \mathbb{H}_N^q, \quad (3.12)$$

and

$$\sum_{k=1}^M |w_k P(\mathbf{y}_k)| \sim \int_{\mathbb{T}^q} |P(\mathbf{t})| d\mu_q^*(\mathbf{t}), \quad P \in \mathbb{H}_N^q, \quad (3.13)$$

where the constants involved in "$\sim$" depend only on q *and not on* $\mathcal{C}$, M, N, *the choice of the weights* w_k, *or the polynomials* P.

Remark Let $\mathcal{C} \subset \mathbb{T}^q$ be a finite set. It is easy to verify that $\eta(\mathcal{C}) \leq 4\delta(\mathcal{C})$. When $\delta(\mathcal{C}) \leq 1$, we can always select a subset of $\mathcal{C}$ for which the minimal separation and the density content have the same order of magnitude. Let m be the integer part of $2\pi/(3\delta(\mathcal{C}))$. For a multi-integer $\mathbf{k}$ with $0 \leq \mathbf{k} \leq m - 1$, we define

$$I_{\mathbf{k}} = I_{\mathbf{k},\mathcal{C}} \overset{\text{def}}{=} \prod_{j=1}^q \left[-\pi + \frac{2k_j\pi}{m}, -\pi + \frac{2(k_j + 1)\pi}{m} \right]. \quad (3.14)$$

Then $\{I_{\mathbf{k}}\}$'s are mutually disjoint arcs except for common boundaries of measure 0 and their union is $\mathbb{T}^q$. Let $\mathbf{z}_k$ be the center of $I_{\mathbf{k}}$. Since each side of the arc $I_{\mathbf{k}}$ is $\geq 3\delta(\mathcal{C})$, the set $\mathcal{C}$ has at least one element in the arc $\{\mathbf{x} : |\mathbf{x} - \mathbf{z}_k|_\infty \leq \delta(\mathcal{C})\} \subset I_{\mathbf{k}}$. We form a subset $\mathcal{C}'$ of $\mathcal{C}$ by choosing exactly one such element for each $\mathbf{k}$. Then it is easy to see that $\delta(\mathcal{C}') \leq 4\delta(\mathcal{C})$, and $\eta(\mathcal{C}') \geq \delta(\mathcal{C}) \geq (1/4)\delta(\mathcal{C}')$. Thus, $(1/4)\delta(\mathcal{C}') \leq \eta(\mathcal{C}') \leq 4\delta(\mathcal{C}')$. In all of our discussion below, we will therefore assume that such a subset $\mathcal{C}'$ has been chosen. The rest of the elements of $\mathcal{C}$ do not make any difference to the statements; e.g., we may just set the weights corresponding the points not so chosen to be 0. Therefore, rather than complicating our notations, we will just identify $\mathcal{C}'$ with $\mathcal{C}$ in our notations. Thus, we assume that

$$(1/4)\delta(\mathcal{C}) \le \eta(\mathcal{C}) \le 4\delta(\mathcal{C}) \le 4, \tag{3.15}$$

and that each $I_{\mathbf{k},\mathcal{C}}$ contains exactly one element of $\mathcal{C}$. Then $M \overset{\text{def}}{=} |\mathcal{C}| = m^q$, and we will re-index $\mathcal{C}$ by setting $\mathbf{y_k}$ to be the unique element of $\mathcal{C} \cap I_{\mathbf{k}}$. $\qquad\square$

One of the important steps in the proof of Theorem 3.5 is the following lemma.

Lemma 3.1 *Let $m \ge 1$ be an integer, let $\{I_{\mathbf{k}} : 0 \le \mathbf{k} \le m-1\}$ be a partition of $\mathbb{T}^q$ as defined in (3.14), $\mathcal{C} \overset{\text{def}}{=} \{\mathbf{y_k}\}_{0 \le \mathbf{k} \le m-1} \subset \mathbb{T}^q$, where each $\mathbf{y_k} \in I_{\mathbf{k}}$, and let (3.15) be satisfied. Then there exists $C > 0$ such that if $\epsilon > 0$ and $N = C\epsilon m$, then we have for every $P \in \mathbb{H}_N^q$,*

$$\left| \frac{1}{m^q} \sum_{0 \le \mathbf{k} \le m-1} |P(\mathbf{y_k})| - \int_{\mathbb{T}^q} |P(\mathbf{z})| \, d\mu_q^*(\mathbf{z}) \right| \le \epsilon \|P\|_1, \tag{3.16}$$

and

$$\left| \frac{1}{m^q} \sum_{0 \le \mathbf{k} \le m-1} P(\mathbf{y_k}) - \int_{\mathbb{T}^q} P(\mathbf{z}) \, d\mu_q^*(\mathbf{z}) \right| \le \epsilon \|P\|_1. \tag{3.17}$$

The lemma is proved in much greater generality in [20]. The ideas behind the proof are quite well known, e.g. [77, Section 4.9.1] or [67]. Here we give a simplified version of the proof in [20], using some ideas in [54,64,65]. Our objective is to highlight the use of localized kernels.

Proof of Lemma 3.1 In this proof, let $\delta = \delta(\mathcal{C})$, and $\eta = \eta(\mathcal{C})$. We note that $\delta(\mathcal{C}) \sim 1/m$. Let $N \ge 1$ be an integer to be chosen later, and $P \in \mathbb{H}_N^q$. Using the mean value theorem, it is easy to see that

$$\max_{\mathbf{z} \in I_{\mathbf{k}}} |P(\mathbf{x}) - P(\mathbf{y_k})| \le \frac{c}{m} \max_{1 \le r \le q} \|\partial_r P\|_{\infty, I_{\mathbf{k}}}. \tag{3.18}$$

Consequently,

$$\left| \frac{1}{m^q} \sum_{0 \le \mathbf{k} \le m-1} |P(\mathbf{y_k})| - \int_{\mathbb{T}^q} |P(\mathbf{z})| \, d\mu_q^*(\mathbf{z}) \right|$$

$$= \left| \sum_{0 \le \mathbf{k} \le m-1} \int_{I_{\mathbf{k}}} (|P(\mathbf{y_k})| - |P(\mathbf{z})|) \, d\mu_q^*(\mathbf{z}) \right|$$

$$\leq \sum_{0 \leq \mathbf{k} \leq m-1} \int_{I_{\mathbf{k}}} |P(\mathbf{y}_{\mathbf{k}}) - P(\mathbf{z})| \, d\mu_q^*(\mathbf{z})$$

$$\leq \max_{1 \leq r \leq q} \frac{c}{m^q} \sum_{0 \leq \mathbf{k} \leq m-1} \|\partial_r P\|_{\infty, I_{\mathbf{k}}}. \tag{3.19}$$

Now, let $S > q$ be an integer, h be an infinitely differentiable low pass filter, $1 \leq r \leq q$, and $H_r(\mathbf{u}) = iu_r h(|\mathbf{u}|_2)$. Then H_r is also infinitely differentiable. Hence, the fact that $|\mathbf{t}|_2 \sim |\mathbf{t}|_\infty$ for all $\mathbf{t} \in \mathbb{R}^q$ and the estimate (3.2) used with H_r in place of H shows that for $\mathbf{t} \in \mathbb{T}^q$,

$$|\partial_r \Phi_N(h, \mathbf{t})| = N|\Phi_N(H_r, \mathbf{t})| \leq c \frac{N^{q+1}}{\max(1, (N|\mathbf{t}|_\infty)^S)}. \tag{3.20}$$

We observe that for $\mathbf{z} \in \mathbb{T}^q$,

$$\partial_r P(\mathbf{z}) = \int_{\mathbb{T}^q} P(\mathbf{t}) \partial_r \Phi_N(h, \mathbf{z} - \mathbf{t}) \, d\mu_q^*(\mathbf{t}),$$

and from this deduce that

$$\sum_{0 \leq \mathbf{k} \leq m-1} \|\partial_r P\|_{\infty, I_{\mathbf{k}}} \leq N \int_{\mathbb{T}^q} |P(\mathbf{t})| \left\{ \sum_{0 \leq \mathbf{k} \leq m-1} \max_{\mathbf{z} \in I_{\mathbf{k}}} |\Phi_N(H_r, \mathbf{z} - \mathbf{t})| \right\} d\mu_q^*(\mathbf{t}).$$
$$\tag{3.21}$$

For the rest of the proof let R_r denote the maximum of the expression in the braces in the above formula for all $\mathbf{t} \in \mathbb{T}^q$. In view of translation invariance of the kernels Φ_N, we may assume without loss of generality that

$$R_r = \sum_{0 \leq \mathbf{k} \leq m-1} \max_{\mathbf{z} \in I_{\mathbf{k}}} |\Phi_N(H_r, \mathbf{z} - (-\pi, \ldots, -\pi))|. \tag{3.22}$$

We conclude using (3.19) that

$$\left| \frac{1}{m^q} \sum_{0 \leq \mathbf{k} \leq m-1} |P(\mathbf{y}_{\mathbf{k}})| - \int_{\mathbb{T}^q} |P(\mathbf{z})| \, d\mu_q^*(\mathbf{z}) \right| \leq c \frac{N}{m^{q+1}} \left(\max_{1 \leq r \leq q} R_r \right) \|P\|_1. \tag{3.23}$$

For the rest of the proof let $\beta = (2\pi N)/m$. For $\ell = 0, 1, \ldots, m - 1$, let

$$\mathcal{J}_\ell = \left\{ \mathbf{k} \in \mathbb{Z}^q : 1 \leq \mathbf{k} \leq m - 1, \frac{2\pi\ell}{m} \leq \min_{\mathbf{z} \in I_{\mathbf{k}}} |\mathbf{z} + (\pi, \ldots, \pi)|_\infty \leq \frac{2\pi(\ell+1)}{m} \right\},$$

and let n_ℓ denote the number of elements in $\mathcal{J}_\ell$. Then $n_\ell \sim \ell^{q-1}$. Therefore, (3.20) shows that

$$R_r \leq cN^q \sum_{\ell=0}^{m-1} \frac{1}{\max(1,(\beta\ell)^S)} n_\ell = cN^q \left\{ \sum_{\ell \leq \beta^{-1}} \ell^{q-1} + \beta^{-S} \sum_{\ell > \beta^{-1}} \ell^{q-1-S} \right\}$$

$$\leq cN^q \beta^{-q} \leq cm^q.$$

Moreover, the very last constant c above can be chosen independently of r. Hence, (3.23) shows that

$$\left| \frac{1}{m^q} \sum_{0 \leq \mathbf{k} \leq m-1} |P(\mathbf{y_k})| - \int_{\mathbb{T}^q} |P(\mathbf{z})| \, d\mu_q^*(\mathbf{z}) \right| \leq c\frac{N}{m}.$$

Choosing $N = \epsilon m/c$, we arrive at (3.16).

Since

$$\left| \frac{1}{m^q} \sum_{0 \leq \mathbf{k} \leq m-1} P(\mathbf{y_k}) - \int_{\mathbb{T}^q} P(\mathbf{z}) \, d\mu_q^*(\mathbf{z}) \right|$$

$$= \left| \sum_{0 \leq \mathbf{k} \leq m-1} \int_{I_\mathbf{k}} (P(\mathbf{y_k}) - P(\mathbf{z})) \, d\mu_q^*(\mathbf{z}) \right|$$

$$\leq \sum_{0 \leq \mathbf{k} \leq m-1} \int_{I_\mathbf{k}} |P(\mathbf{y_k}) - P(\mathbf{z})| \, d\mu_q^*(\mathbf{z})$$

$$\leq \max_{1 \leq r \leq q} \frac{c}{m^{q+1}} \sum_{0 \leq \mathbf{k} \leq m-1} \|\partial_r P\|_{\infty, I_\mathbf{k}},$$

the same proof as above shows also that (3.17) holds as well.   □

The other important ingredient in the proof of Theorem 3.5 is the following consequence of the Hahn–Banach theorem, known as the Krein extension theorem; see [21] for a recent proof. Let $\mathbb{X}$ be a normed linear space, $\mathcal{K}$ be a subset of its normed dual $\mathbb{X}^*$, and $\mathcal{V}$ be a linear subspace of $\mathbb{X}$. We say that a linear functional $x^* \in \mathcal{V}^*$ is positive on $\mathcal{V}$ with respect to $\mathcal{K}$ if $x^*(f) \geq 0$ for every $f \in \mathcal{V}$ with the property that $y^*(f) \geq 0$ for every $y^* \in \mathcal{K}$.

Theorem 3.6 *Let $\mathbb{X}$ be a normed linear space, let $\mathcal{K}$ be a bounded subset of its normed dual $\mathbb{X}^*$, let $\mathcal{V}$ be a linear subspace of $\mathbb{X}$, and let $x^* \in \mathcal{V}^*$ be positive on $\mathcal{V}$ with respect to $\mathcal{K}$. We assume further that there exists $v_0 \in \mathcal{V}$ such that $\|v_0\|_{\mathbb{X}} = 1$ and*

$$\inf_{y^* \in \mathcal{K}} y^*(v_0) = \beta^{-1} > 0. \tag{3.24}$$

Then there exists an extension $X^* \in \mathbb{X}^*$ of x^* which is positive on $\mathbb{X}$ with respect to $\mathcal{K}$ and satisfies

$$\|X^*\|_{\mathbb{X}^*} \leq \beta \sup_{y^* \in \mathcal{K}} \|y^*\|_{\mathbb{X}*} x^*(v_0). \qquad (3.25)$$

With this preparation, we are ready to prove Theorem 3.5.

Proof of Theorem 3.5 As explained earlier, we may assume that $M = m^q$ and $\mathcal{C} = \{\mathbf{y_k} : 0 \leq \mathbf{k} \leq m - 1\}$.

We take $\epsilon = 1/4$ in Lemma 3.1 and conclude from (3.16) that

$$(3/4)\|P\|_1 \leq \frac{1}{m^q} \sum_{0 \leq \mathbf{k} \leq m-1} |P(\mathbf{y_k})| \leq (5/4)\|P\|_1, \qquad (3.26)$$

and from (3.17) that

$$\left| \frac{1}{m^q} \sum_{0 \leq \mathbf{k} \leq m-1} P(\mathbf{y_k}) - \int_{\mathbb{T}^q} P(\mathbf{z}) \, d\mu_q^*(\mathbf{z}) \right| \leq (1/4)\|P\|_1 \leq \frac{1}{3m^q} \sum_{0 \leq \mathbf{k} \leq m-1} |P(\mathbf{y_k})|. \qquad (3.27)$$

Denote by $\mathbb{X}$ the space $\mathbb{R}^M$, equipped with the norm

$$\|\mathbf{x}\| = \sum_{0 \leq \mathbf{k} \leq m-1} \mu_q^*(I_\mathbf{k})|x_\mathbf{k}| \quad \text{where} \quad \mathbf{x} = (x_\mathbf{k})_{0 \leq \mathbf{k} \leq m-1}.$$

For the set $\mathcal{K}$, we choose the set of coordinate functionals; $z_\mathbf{k}^*(\mathbf{x}) = x_\mathbf{k}$. Then $\mathcal{K}$ is clearly a compact subset of $\mathbb{X}^*$. We consider the operator $\mathcal{S} : \mathbb{H}_N^q \to \mathbb{R}^M$ given by $P \mapsto (P(\mathbf{y_k}))_{\mathbf{k}=0}^{m-1}$, and take the subspace $\mathcal{V}$ of $\mathbb{X}$ to be the range of $\mathcal{S}$. The lower estimate in (3.26) shows that $\mathcal{S}$ is invertible on $\mathcal{V}$. We define the functional x^* on $\mathcal{V}$ by

$$x^*(\mathcal{S}(P)) = \int_{\mathbb{T}^q} P(\mathbf{z}) \, d\mu_q^*(\mathbf{z}) - \frac{1}{3m^q} \sum_{0 \leq \mathbf{k} \leq m-1} P(\mathbf{y_k}), \quad P \in \mathbb{H}_N^q.$$

Moreover, if each $P(\mathbf{y_k}) \geq 0$, then, using (3.27), we can conclude that

$$\int_{\mathbb{T}^q} P(\mathbf{z}) \, d\mu_q^*(\mathbf{z}) - \frac{1}{3m^q} \sum_{0 \leq \mathbf{k} \leq m-1} P(\mathbf{y_k}) \geq \frac{1}{3m^q} \sum_{0 \leq \mathbf{k} \leq m-1} P(\mathbf{y_k}) \geq 0.$$

Thus, x^* is positive on $\mathcal{V}$ with respect to $\mathcal{K}$. The element $(1, \ldots, 1) \in \mathcal{V}$ serves as v_0 in Theorem 3.6. Theorem 3.6 then implies that there exists a nonnegative functional

X^* on $\mathbb{R}^M$ that extends x^*. We may identify this functional with $(\tilde{W}_{\mathbf{k}})_{\mathbf{k}=0}^{m-1} \in \mathbb{R}^M$ such that each $\tilde{W}_{\mathbf{k}} \geq 0$. The fact that X^* extends x^* means that for each $P \in \mathbb{H}_N^q$,

$$\int_{\mathbb{T}^q} P(\mathbf{z})\, d\mu_q^*(\mathbf{z}) = \sum_{0 \leq \mathbf{k} \leq m-1} (\tilde{W}_{\mathbf{k}} + (1/(3m^q)))P(\mathbf{y}_{\mathbf{k}})$$

$$= \sum_{0 \leq \mathbf{k} \leq m-1} w_{\mathbf{k}} P(\mathbf{y}_{\mathbf{k}}) \quad \text{where} \quad w_{\mathbf{k}} \stackrel{\text{def}}{=} \tilde{W}_{\mathbf{k}} + (1/(3m^q)). \tag{3.28}$$

If we revert to the original set of points and set $w_{\mathbf{k}} = 0$ if $\mathbf{y}_{\mathbf{k}}$ is not in the subset chosen as in the remark following the statement of Theorem 3.5, then this is (3.12).

It was proved in [21, Theorem 5.8] that (3.12) implies (3.13). $\qquad\qquad\square$

Next, we make a few comments about the numerical computation of the weights $w_{\mathbf{k}}$. We observe that the reduction of the original data set $\{\mathbf{y}_k\}$ may be done is several different ways and the weights $w_{\mathbf{k}}$ are not uniquely defined either. Having made the reduction of the original set, a straightforward way to find the weights numerically is the following, see [34]. We minimize $\sum_{0 \leq \mathbf{k} \leq m-1} w_{\mathbf{k}}^2$ subject to the conditions

$$\sum_{0 \leq \mathbf{k} \leq m-1} w_{\mathbf{k}} \exp(i\mathbf{j} \cdot \mathbf{y}_{\mathbf{k}}) = \begin{cases} 1, & \text{if } j = 0, \\ 0, & \text{otherwise,} \end{cases}$$

for $0 \leq \mathbf{j} < N$. This involves the solution of a linear system of equations whose matrix, the so-called *Gram matrix*, is given by

$$V_{\mathbf{j},\ell} = \sum_{0 \leq \mathbf{k} \leq m-1} \exp(i(\mathbf{j} - \boldsymbol{\ell}) \cdot \mathbf{y}_{\mathbf{k}}), \quad 0 \leq \mathbf{j}, \boldsymbol{\ell} < N. \tag{3.29}$$

If $(a_j)_{0 \leq \mathbf{j} < N}$ is an arbitrary vector and we take $P(\mathbf{x}) = \sum_{0 \leq \mathbf{j} < N} a_{\mathbf{j}} \exp(i\mathbf{j} \cdot \mathbf{x})$, then the Raleigh quotient for this matrix can be calculated to be

$$\frac{\sum_{0 \leq \mathbf{k} \leq m-1} |P(\mathbf{y}_{\mathbf{k}})|^2}{\sum_{0 \leq \mathbf{j} < N} |a_{\mathbf{j}}|^2} = \frac{\sum_{0 \leq \mathbf{k} \leq m-1} |P(\mathbf{y}_{\mathbf{k}})|^2}{\|P\|_2^2},$$

where we used the Parseval identity in the last step. Thus, the largest and smallest eigenvalues of V, $\lambda_{\max}$ and $\lambda_{\min}$ are given by

$$\lambda_{\min} = \min_{P \in \mathbb{H}_N^q} \frac{\sum_{0 \leq \mathbf{k} \leq m-1} |P(\mathbf{y}_{\mathbf{k}})|^2}{\|P\|_2^2}, \quad \lambda_{\max} = \max_{P \in \mathbb{H}_N^q} \frac{\sum_{0 \leq \mathbf{k} \leq m-1} |P(\mathbf{y}_{\mathbf{k}})|^2}{\|P\|_2^2}, \tag{3.30}$$

see [29, Theorem 4.2.2, p. 176]. In practice, one has to choose N by trial and error so that the condition number of V is "reasonable". We are tempted to solve a non-linear optimization problem with the additional requirement that the weights should be non-negative. It is our experience that the non-negativity of the weights is not important in practice, but an inequality of the form (3.13) is essential. For any given data, such an inequality will, of course, hold with some constants depending on N and the data set. However, it is of interest to estimate the constants. It turns out that the constants involved are proportional to $\lambda_{\max}$ and $\lambda_{\min}$.

We would like to point out another interesting fact. Suppose one wishes to find a least squares fit from $\mathbb{H}_N^q$ to the data of the form $\{(\mathbf{y_k}, z_\mathbf{k})\}_{0 \leq \mathbf{k} \leq m-1}$, that is, find $a_\mathbf{j}$'s to minimize

$$\sum_{0 \leq \mathbf{k} \leq m-1} |w_\mathbf{k}| \left| z_\mathbf{k} - \sum_{0 \leq \ell < N} a_\ell \exp(i\ell \cdot \mathbf{y_k}) \right|^2$$

for a suitable choice of $w_\mathbf{k}$. This involves the solution of a linear system of equations where the matrix involved is G, defined by

$$G_{\mathbf{j},\ell} = \sum_{0 \leq \mathbf{k} \leq m-1} |w_\mathbf{k}| \exp(i(\mathbf{j} - \ell) \cdot \mathbf{y_k}).$$

As before, we have the bounds

$$\tilde{\lambda}_{\min} \sum_{0 \leq \mathbf{k} \leq m-1} |w_\mathbf{k}| |P(\mathbf{y_k})|^2 \leq \|P\|_2^2 \leq \tilde{\lambda}_{\max} \sum_{0 \leq \mathbf{k} \leq m-1} |w_\mathbf{k}| |P(\mathbf{y_k})|^2, \quad P \in \mathbb{H}_N^q,$$

$$(3.31)$$

where $\tilde{\lambda}_{\min}$ and $\tilde{\lambda}_{\max}$ are the smallest and largest eigenvalues of G.

To describe these results while keeping track of the constants on the various data sets and weights, etc., and also to simplify our notation in further theory, it is convenient to use a measure notation. The actual choice of the reduced data set and the weights plays no role in our theoretical consideration. Therefore, it is convenient to define a measure ν that associates the mass $w_\mathbf{k}$ with each $\mathbf{y_k}$, that is, for any subset $B \subset \mathbb{T}^q$,

$$\nu(B) = \sum_{\mathbf{k}:\mathbf{y_k} \in B} w_\mathbf{k}.$$

We pause in our discussion to review some basic notions related to signed measures and introduce some notation before proceeding further. We recall that the total variation measure of any signed measure μ is defined by

$$|\mu|(\mathcal{U}) \overset{\text{def}}{=} \sup \sum_{i=1}^{\infty} |\mu(U_i)|, \quad \mathcal{U} \subset \mathbb{T}^q,$$

where the supremum is taken over all countable partitions $\{U_i\}$ into measurable sets of $\mathcal{U}$. For the measure ν as defined above, one can easily deduce that $|\nu|(B) = \sum_{\mathbf{k}:\mathbf{y_k} \in B} |w_\mathbf{k}|$ for any subset $B \subset \mathbb{T}^q$. It is well known that any signed measure μ on $\mathbb{T}^q$ satisfies $|\mu|(\mathbb{T}^q) < \infty$. The support of a measure μ, denoted by $\mathsf{supp}(\mu)$, is the set of all $\mathbf{x} \in \mathbb{T}^q$ such that for every open subset U of $\mathbb{T}^q$ containing $\mathbf{x}$, $|\mu|(U) > 0$.

If $1 \le p \le \infty$, μ is a (possibly signed) measure on $\mathbb{T}^q$, $B \subset \mathbb{T}^q$ is μ-measurable, and $f : B \to \mathbb{C}$ is μ-measurable, then the L^p norm of f with respect to μ is given by

$$\|f\|_{\mu;p,B} \stackrel{\text{def}}{=} \begin{cases} \left\{ \int_B |f(x)|^p \, d\,|\mu| \right\}^{1/p}, & \text{if } 1 \le \mathrm{p} < \infty, \\ |\mu| - \operatorname{ess\,sup} |f(x)|_{x \in B}, & \text{if } \mathrm{p} = \infty. \end{cases} \qquad (3.32)$$

As before, we will omit the mention of B if $B = \mathbb{T}^q$, and to keep the notation consistent, will also omit the measure μ from the notation if $\mu = \mu_q^*$. The space $X^p(\mu)$ denotes the $L^p(\mu)$ closure of the set of all trigonometric polynomials. In this notation, the estimate (3.13) becomes

$$\|P\|_{\nu;1} \sim \|P\|_1, \quad P \in \mathbb{H}_N^q,$$

and (3.31) can be written as

$$\tilde{\lambda}_{\min} \|P\|_2^2 \le \|P\|_{\nu;2}^2 \le \tilde{\lambda}_{\max} \|P\|_2^2.$$

In the sequel, we will not fix the measure ν any more as above, and instead we use the notations ν, μ, etc., to denote arbitrary measures. With a different measure ν, the formulas (3.30) become

$$\lambda_{\min} \|P\|_2^2 \le \|P\|_{\nu;2}^2 \le \lambda_{\max} \|P\|_2^2, \quad P \in \mathbb{H}_N^q.$$

We are now ready to resume our discussion of the M–Z inequalities and related topics.

Definition 3.3 (a) A (possibly signed) measure μ is called a **quadrature measure** of order n when

$$\int_{\mathbb{T}^q} T \, d\mu = \int_{\mathbb{T}^q} T \, d\mu_q^*, \quad T \in \mathbb{H}_n^q. \qquad (3.33)$$

(b) A (possibly signed) measure μ is called a **Marcinkiewicz–Zygmund measure**, or M–Z measure, of order n when the following M–Z inequality

$$\int_{\mathbb{T}^q} |T| \, d|\mu| = \|T\|_{\mu;1} \le c(n,\mu) \|T\|_1, \quad T \in \mathbb{H}_n^q, \qquad (3.34)$$

is satisfied where $c(n, \mu)$ is a constant independent of T. The smallest c that works in (3.34) will be denote by $\|\mu\|_n$.

It can be shown easily that for each $n \in \mathbb{N}$, $\| \circ \|_n$ is a norm on the space of Radon measures. Clearly, for every $n \in \mathbb{N}$, μ_q^* itself is an M–Z quadrature measure of order n with $\|\mu_q^*\|_n = 1$. In general, if μ is an M–Z quadrature measure of order n for some $n > 0$, then $\|\mu\|_n \ge 1$.

In [21, Proposition 2.1, Theorem 5.4], we proved the following.

Theorem 3.7 *Let μ be a Radon measure and let $n \geq 2$. Then we have the following inequalities where the constants c are independent of both μ and n.*

(a) *If $1 \leq p < \infty$ then*

$$\|P\|_{\mu;p} \leq c\|\mu\|_n^{1/p}\|P\|_p, \quad P \in \mathbb{H}_n^q. \tag{3.35}$$

Conversely, if for some $p \in [1, \infty)$ and $A = A(\mu, n) > 0$,

$$\|P\|_{\mu;p} \leq cA^{1/p}\|P\|_p, \quad P \in \mathbb{H}_n^q,$$

holds, then $\|\mu\|_n \leq cA$.

(b) *For any $r > 0$ and $\mathbf{x} \in \mathbb{T}^q$ we have $|\mu|(\{\mathbf{y} : \|\mathbf{x} - \mathbf{y}\|_\infty \leq r\}) \leq c\|\mu\|_n(r + 1/n)^q$. Conversely, if there exists a constant $A = A(\mu, n)$ such that for every $r > 0$ and $\mathbf{x} \in \mathbb{T}^q$, $|\mu|(\{\mathbf{y} : \|\mathbf{x} - \mathbf{y}\|_\infty \leq r\}) \leq A(r + 1/n)^q$, then $A \leq c\|\mu\|_n$.*

(c) *For any constant $\alpha > 0$, we have $\|\mu\|_n \sim \|\mu\|_{\alpha n}$, where the constants involved in the $\sim$ relationship depend only on α (and q) but not on μ or n.*

Once we compute the maximum eigenvalue of the Gram matrix V defined in (3.29), Theorem 3.7(a) allows us to estimate the constant involved in (3.13). Theorem 3.7(b) gives a geometric criteria for (3.13) without referring to trigonometric polynomials. Theorem 3.7(c) shows that if (3.13) holds for some n, then it holds also with equivalent constants for αn for every $\alpha > 1$ as well. In particular, even if a quadrature measure of order n supported on the minimal number of points, $\dim(\mathbb{H}_{n/2}^q)$, does not exist when $q \geq 2$, M–Z measures of order n with this property are plentiful.

3.4 Wavelet-like representation

Our starting point here is the following analogue of Theorem 3.1, where the novelty is that in the case of functions in X^∞, their samples are used for approximation.

First, some notation. If μ is a (possibly signed) measure and $f \in L^1(\mu)$, we define

$$\hat{f}(\mu; \mathbf{k}) = \int_{\mathbb{T}^q} f(\mathbf{t})\exp(-i\mathbf{k} \cdot \mathbf{t})\,d\mu(\mathbf{t}), \quad \mathbf{k} \in \mathbb{Z}^q.$$

If $\mu = \mu_q^*$ then $\hat{f}(\mu; \mathbf{k}) = \hat{f}(\mathbf{k})$. If μ is discretely supported measure then $\hat{f}(\mu; \mathbf{k})$ is a discretized approximation to $\hat{f}(\mathbf{k})$ as indicated by μ. A common example is the discrete Fourier transform, obtained by letting μ be supported on a set of dyadic points on $\mathbb{T}^q$. It is well known that the Fast Fourier Transform (FFT) algorithm allows a fast computation of the coefficients $\hat{f}(\mu; \mathbf{k})$ for the appropriate measures μ, and, for this reason, it is used widely in engineering applications. In the case when μ is sparsely supported, there are new algorithms with sublinear complexity, see [28]. There are many other examples, including in particular, the so-called low discrepancy quadrature rules, see, e.g., [13] for a detailed discussion and further references. Our interest here is in the case when μ is an M–Z quadrature measure.

The analogue of the summability operator is defined by

$$
\begin{aligned}
\sigma_n(\mu; H, f, \mathbf{x}) &\stackrel{\text{def}}{=} \sum_{0 \leq \mathbf{k} < n} H\left(\frac{\mathbf{k}}{n}\right) \hat{f}(\mu; \mathbf{k}) \exp(i\mathbf{k} \cdot \mathbf{x}) \\
&= \int_{\mathbb{T}^q} f(\mathbf{t}) \Phi_n(H; \mathbf{x} - \mathbf{t}) \, d\mu(\mathbf{t}), \quad n \in \mathbb{N}, \ \mathbf{x} \in \mathbb{T}^q, \ f \in L^1(\mu).
\end{aligned}
\tag{3.36}
$$

We will adopt the same conventions as in Sect. 3.2 with respect to overloaded notations. For example, when $H(\mathbf{t}) = h(|\mathbf{t}|_2)$, we will write $\sigma_n(\mu; h, f)$ in place of $\sigma_n(\mu; H, f)$.

The analogue of Theorem 3.1 is the following.

Theorem 3.8 *Let $n \in \mathbb{N}$, let $S > q$ be an integer, $1 \leq p \leq \infty$, and let μ be an M–Z quadrature measure of order $3n/2$. Let h be an S-times continuously differentiable low pass filter.*

(a) *For for all $P \in \mathbb{H}^q_{n/2}$, we have $\sigma_n(\mu; h, P) = P$.*

(b) *We have for all $f \in X^p(\mu)$,*

$$
\|\sigma_n(\mu; h, f)\|_p \leq c \|\mu\|_n^{1-1/p} \|f\|_{\mu; p}.
\tag{3.37}
$$

Consequently, if $f \in X^\infty$, then $\|\sigma_n(\mu; h, f)\|_\infty \leq c\|\mu\|_n \|f\|_\infty$, and, furthermore,

$$
E_{n,\infty}(f) \leq \|f - \sigma_n(\mu; h, f)\|_\infty \leq c\|\mu\|_n E_{n/2,\infty}(f).
\tag{3.38}
$$

(c) *If $f \in L^1(\mu)$ is supported on a compact set K and V is an open set with $K \subset V$, then*

$$
\|\sigma_n(\mu; h, f)\|_{\infty, \mathbb{T}^q \setminus V} \leq c\|f\|_{\mu;1} n^{q-S},
\tag{3.39}
$$

where, in addition to S and h, the constant c may depend on K and V.

Proof If $P \in \mathbb{H}^q_{n/2}$, then for each $\mathbf{x} \in \mathbb{T}^q$ we have $P\Phi_n(h, \mathbf{x} - \circ) \in \mathbb{H}^q_{3n/2}$. Since μ is a quadrature measure of order $3n/2$,

$$
P(\mathbf{x}) = \int_{\mathbb{T}^q} P(\mathbf{t})\Phi_n(h, \mathbf{x} - \mathbf{t}) \, d\mu_q^*(\mathbf{t}) = \int_{\mathbb{T}^q} P(\mathbf{t})\Phi_n(h, \mathbf{x} - \mathbf{t}) \, d\mu(\mathbf{t}) = \sigma_n(\mu; h, P, \mathbf{x}).
$$

This proves part (a). The proof of part (b) is almost verbatim the same as that of Theorem 2.1(b), except that multivariate notation needs to be used, and, in addition, we need the fact that μ is an M–Z measure of order $3n/2$ to conclude that for each $\mathbf{x} \in \mathbb{T}^q$

$$\int_{\mathbb{T}^q} |\Phi_n(h, \mathbf{x} - \mathbf{t})| d|\mu|(\mathbf{t}) \le c\|\mu\|_n \int_{\mathbb{T}^q} |\Phi_n(h, \mathbf{x} - \mathbf{t})| d\mu_q^*(\mathbf{t}) \le c\|\mu\|_n.$$

Part (c) is easy to prove using the localization estimate (3.2). $\qquad\square$

As an immediate corollary, we note the following complement to the one-sided inequality (3.34), cf. Lemma 2.1 and (3.31).

Corollary 3.2 *Let $n \in \mathbb{N}$ and let μ be an M–Z quadrature measure of order $3n$. Then for $P \in \mathbb{H}_n^q$ and $1 \le p < \infty$,*

$$\|P\|_p \le c\|\mu\|_n^{1-1/p}\|P\|_{\mu;p} \le c_1\|\mu\|_n\|P\|_p. \tag{3.40}$$

Proof We use (3.37) with P in place of f and σ_{2n} in place of σ_n. Since $\sigma_{2n}(\mu; h, P) = P$, this leads to the first inequality in (3.40). The second inequality is a consequence of Theorem 3.7(a). $\qquad\square$

Next, we come to the definition and characterization of local smoothness classes. The local smoothness classes are defined exactly as in Definition 2.3, except that multivariate analogues of $\mathbb{T}$, "arc", and smoothness classes are used. Our objective is to state the analogues of Theorems 2.5 and 2.6. Since frames in the sense of L^2 cannot be defined using values of the target function at countably many points, and the topic of tight frames does not add anything new to our discussion, we will not delve into the topic of tight frames.

We say a sequence of measures $\boldsymbol{\mu} = \{\mu_n\}$ has nested support when

$$m < n \implies \mathsf{supp}(\mu_m) \subseteq \mathsf{supp}(\mu_n).$$

In theoretical considerations for approximation based on values of the target function, we will usually assume that the data is available as a sequence $\{\mathcal{C}_m\}$ of finite subsets of $\mathbb{T}^q$. By taking unions, we may assume without loss of generality that the sets are nested. The following proposition, proved in [49, Proposition 2.1], shows that the data reduction and construction of M–Z quadrature formulas c an be done in a consistent manner.

Proposition 3.1 *Let $\{\mathcal{C}_m\}$ be a sequence of finite subsets of $\mathbb{T}^q$ with $\delta(\mathcal{C}_m) \sim 1/m$, and let $\mathcal{C}_m \subseteq \mathcal{C}_{m+1}$ for $m \in \mathbb{N}$. Then there exists a sequence of subsets $\{\tilde{\mathcal{C}}_m \subseteq \mathcal{C}_m\}$, where, for $m \in \mathbb{N}$, we have $\delta(\tilde{\mathcal{C}}_m) \sim 1/m$, $\tilde{\mathcal{C}}_m \subseteq \tilde{\mathcal{C}}_{m+1}$, and $\delta(\tilde{\mathcal{C}}_m) \le 2\eta(\tilde{\mathcal{C}}_m)$.*

Definition 3.4 *Let $h : \mathbb{R} \to [0, \infty)$ be compactly supported and let $\boldsymbol{\mu} = \{\mu_n\}_{n=0}^\infty$ be a sequence of Borel measures on $\mathbb{T}^q$ with nested support. Let $f \in \cap_{n=0}^\infty L^1(\mu_n)$. Then, for each non-negative integer n, the band-pass operator $\tau_n(\boldsymbol{\mu}; h, f)$ with respect to $\boldsymbol{\mu}$*

is given by

$$\tau_0(\pmb{\mu}; h, f) \stackrel{\text{def}}{=} \sigma_1(\mu_0; h, f) \text{ and } \tau_n(\pmb{\mu}; h, f)$$

$$\stackrel{\text{def}}{=} \sigma_{2^n}(\mu_n; h, f) - \sigma_{2^{n-1}}(\mu_{n-1}; h, f), \text{ for } n \in \mathbb{N}. \quad (3.41)$$

To keep the notation consistent, we will generally omit the symbol $\pmb{\mu}$ if each $\mu_n = \mu_q^*$. If we must mention this measure for emphasis, then we will write $\tau_j(\mu_q^*; h, f)$ in this case.

The kernels Ψ_j are defined in the same way as (2.27), except for obvious multivariate substitutions.

The analogue of Theorem 2.5 is the following statement.

Theorem 3.9 *Let* $1 \le p \le \infty, \gamma > 0, f \in X^p$, *and let* h *be an infinitely differentiable low pass filter. Let* $\pmb{\nu} = \{\nu_j\}_{j=0}^{\infty}$ *be a sequence of measures such that each* ν_j *is a quadrature measure of order* 2^{j+1}. *Further, let* $\pmb{\mu} = \{\mu_j\}_{j=0}^{\infty}$ *be a sequence of measures such that each* μ_j *is an M–Z quadrature measure of order* $3 \times 2^{j-1}$.

(a) *We have*

$$f = \sum_{j=0}^{\infty} \tau_j(h, f) = \sum_{j=0}^{\infty} \int_{\mathbb{T}^q} \tau_j(h, f, \mathbf{t}) \Psi_j(h, \circ - \mathbf{t}) \, d\nu_j(\mathbf{t})$$

with convergence in the sense of X^p.

(b) *If* $f \in L^2$ *then*

$$\|f\|_2^2 \le \sum_{j=0}^{\infty} \|\tau_j(h, f)\|_2^2 = \sum_{j=0}^{\infty} \int_{\mathbb{T}^q} |\tau_j(h, f, \mathbf{t})|^2 \, d\nu_j(\mathbf{t}) \le 4\|f\|_2^2. \quad (3.42)$$

(c) *If* $f \in X^{\infty}$, *then*

$$f = \sum_{j=0}^{\infty} \tau_j(\pmb{\mu}; h, f) = \sum_{j=0}^{\infty} \int_{\mathbb{T}^q} \tau_j(\mu_j; h, f, \mathbf{t}) \Psi_j(h, \circ - \mathbf{t}) \, d\nu_j(\mathbf{t}), \quad (3.43)$$

where both the series converge uniformly.

The interest in the above theorem is clearly when $\pmb{\nu}$ is a sequence of discretely supported measures. The proofs of this theorem and of the following analogue of Theorem 2.6 are verbatim the same as those of their univariate analogues, and therefore we omit them.

Theorem 3.10 *Let* $1 \leq p \leq \infty$, $\gamma > 0$, $\mathbf{x}_0 \in \mathbb{T}^q$, $f \in X^p$, *and let h be an infinitely differentiable low pass filter. Let* $\boldsymbol{\mu}$ *and* $\boldsymbol{\nu}$ *be sequences of measures as in Theorem 3.9. We assume further that* $\sup_{j \geq 0} \|\mu_j\|_{2^j} < \infty$ *and* $\sup_{j \geq 0} \|\nu_j\|_{2^j} < \infty$. *Then the following are equivalent.*

(a) $f \in W_{\gamma, p}(\mathbf{x}_0)$.

(b) *There exists an arc* $I \ni \mathbf{x}_0$ *such that*

$$\|\tau_j(h, f)\|_{\nu_j; p, I} = \mathcal{O}(2^{-j\gamma}). \tag{3.44}$$

(c) *In the case when* $p = \infty$,

$$\max_{\mathbf{t} \in \mathrm{supp}(\nu_j) \cap I} |\tau_j(\boldsymbol{\mu}; h, f, \mathbf{t})| = \mathcal{O}(2^{-j\gamma}). \tag{3.45}$$

Remark As before, the main interest is when $\boldsymbol{\nu}$ is a sequence of discretely supported measures with nested supports. We note that the polynomials $\tau_j(h, f)$ in the estimates (3.44) are computed using the Fourier coefficients of f. The estimates (3.45) are more general in the sense that the sequence $\boldsymbol{\mu}$ can be a sequence of discrete measures as well, in which case, the values of f at the points in the supports of these measures are used. Of course, such a generalization is possible only if $p = \infty$, so that point evaluations are well defined. $\square$

4 Periodic basis function (PBF) networks

4.1 Trigonometric polynomials and PBFs

We recall that a PBF network is a function of the form $\mathbf{x} \mapsto \sum_{k=1}^{N} a_k G(\mathbf{x} - \mathbf{x}_k)$, $\mathbf{x} \in \mathbb{T}^q$ where $G \in X^\infty$ is called the *activation function*, N is the number of *neurons*, and $\mathbf{x}_k \in \mathbb{T}^q$ are called *centers*. Thus, a PBF network is a translation network with activation function defined on $\mathbb{T}^q$ and the centers in $\mathbb{T}^q$ as well.

Starting with [51], we discovered a close connection between approximation by trigonometric polynomials and that by periodic basis function networks. While all the research known to us prior to [51] gave the degree of approximation by RBF networks in terms of a scaling parameter, the results in [51] appear to be the first of their kind where the theory was developed in terms of the number of neurons. Later on, it was observed by Schaback and Wendland [73], and independently in [42]—in the context of the so-called Gaussian networks—that it is the minimal separation among the centers, cf. Definition 3.2, of the network which leads to a complete theory of direct and converse theorems for approximation in this context. Accordingly, we will formulate our theorems here in terms of the minimal separation among the centers.

In the sequel, let $G \in X^\infty(\mathbb{T}^q)$ be a fixed activation function satisfying the condition that $\hat{G}(\mathbf{k}) \neq 0$ for any $\mathbf{k} \in \mathbb{Z}^q$. For $f \in L^1$, we may then define formally a "derivative" by the formula

$$\widehat{D_G(f)}(\mathbf{k}) \stackrel{\text{def}}{=} \frac{\hat{f}(\mathbf{k})}{\hat{G}(\mathbf{k})}, \quad \mathbf{k} \in \mathbb{Z}^q. \tag{4.1}$$

In the case when $D_G(f) \in L^2$, one says that f is in the native space of G. However, we will not need this concept. For our purpose, it is enough to note that if $P \in \mathbb{H}_n^q$ for some $n \in \mathbb{N}$, then $D_G P \in \mathbb{H}_n^q$ as well. In fact, we verify the following important observation:

Proposition 4.1 *Let* $n \in \mathbb{N}$, $P \in \mathbb{H}_n^q$. *Then*

$$P(\mathbf{x}) = \int_{\mathbb{T}^q} G(\mathbf{x} - \mathbf{y}) D_G(P, \mathbf{y}) \, d\mu_q^*(\mathbf{y}), \quad \mathbf{x} \in \mathbb{T}^q. \tag{4.2}$$

The proof of this proposition is a simple comparison of Fourier coefficients of both sides of (4.2). The basic idea in our proofs of the direct theorems as well as in wavelet-like representations is to discretize the integral in (4.2). The resulting estimate is shown in Theorem 4.1 below. In the remainder of this section, we use the notation

$$\mathfrak{m}_n = \mathfrak{m}_n(G) = \min_{|\mathbf{k}|_2 \le n} |\hat{G}(\mathbf{k})|, \tag{4.3}$$

and

$$x_+ = \max(x, 0), \quad x \in \mathbb{R}.$$

Theorem 4.1 *Let* $1 \le p \le \infty$, $n \in \mathbb{N}$, $N \in \mathbb{N}$, $P \in \mathbb{H}_n^q$, *and let* ν *be an M–Z quadrature measure of order* $n + N$. *Then*

$$\left\| P - \int_{\mathbb{T}^q} G(\circ - \mathbf{y}) D_G(P, \mathbf{y}) \, d\nu(\mathbf{y}) \right\|_\infty \le c \|\nu\|_{n+N} \left(n^{\left(\frac{q}{p} - \frac{q}{2}\right)_+} \right) \frac{E_{N,\infty}(G)}{\mathfrak{m}_n} \|P\|_p. \tag{4.4}$$

We observe that in view of Theorem 3.5, an arbitrary finite subset with sufficiently high density content admits an M–Z quadrature measure of order $n + N$ supported on this subset. In particular, such a measure ν exists with $|\text{supp}(\nu)| \sim (n + N)^q$, $\|\nu\|_{n+N} \le c$, and $\eta(\text{supp}(\nu)) \sim (n + N)^{-1}$. Using such a choice of measure for ν, the integral expression in (4.4) is a PBF network with $\sim (n + N)^q$ neurons, the set of centers being $\text{supp}(\nu)$, and the minimal separation among its centers is $\sim (n+N)^{-1}$.

We note a corollary of Theorem 4.1.

Corollary 4.1 *Let* $1 \le p \le \infty$, *and let* G *satisfy* $\hat{G}(\mathbf{k}) \ne 0$ *for any* $\mathbf{k} \in \mathbb{Z}^q$. *Then the class of all translation networks with* G *as the activation function is dense in* X^p.

Proof If $f \in X^p$ and $\epsilon > 0$, we find $n \in \mathbb{N}$ and $P \in \mathbb{H}_n^q$ such that $\|f - P\|_p < \epsilon/2$. For this n, we may find N large enough so that the network $\mathcal{G}$ as constructed in (4.4) satisfies $\|P - \mathcal{G}\|_p < \epsilon/2$. Then $\|f - \mathcal{G}\|_p < \epsilon$. $\qquad\square$

The estimate (4.4) is much more meaningful in the case when G is very smooth, that is, for every $L > 0$,

$$\lim_{|\mathbf{k}|_2 \to \infty} |\mathbf{k}|_2^L |\hat{G}(\mathbf{k})| = 0.$$

We introduce the following definition.

Definition 4.1 Let $A > 0$. We will say that $G \in \mathcal{E}_A$ if each of the following conditions is satisfied:

1. $G \in X^\infty$,
2. for all $\mathbf{k} \in \mathbb{Z}^q$, $\hat{G}(\mathbf{k}) \neq 0$,
3.

$$\limsup_{n \to \infty} \left(\frac{E_{An,\infty}(G)}{\mathfrak{m}_n} \right)^{1/n} < 1. \tag{4.5}$$

For $G \in \mathcal{E}_A$, Theorem 4.1 (used with $N = An$) leads to the following corollary.

Corollary 4.2 *Let $A > 0$ and $G \in \mathcal{E}_A$. Let $1 \le p \le \infty$, $n \in \mathbb{N}$, $P \in \mathbb{H}_n^q$, and let ν be an M–Z quadrature measure of order $(1 + A)n$. Then there exists $\rho = \rho(A, p)$ such that $0 < \rho < 1$, and*

$$\left\| P - \int_{\mathbb{T}^q} G(\circ - \mathbf{y}) D_G(P, \mathbf{y}) \, d\nu(\mathbf{y}) \right\|_\infty \le c \|\nu\|_n \rho^n \|P\|_p. \tag{4.6}$$

Many standard examples used in network expansions involve activation functions in $\mathcal{E}_A$ for some $A > 0$, cf. [75] for the computation of Fourier transforms given in the examples below.

Example 4.1 Periodization of the Gaussian.

$$G(\mathbf{x}) = \sum_{\mathbf{k} \in \mathbb{Z}^q} \exp(-|\mathbf{x} - 2\pi \mathbf{k}|_2^2/2), \ \hat{G}(\mathbf{k}) = (2\pi)^{q/2} \exp(-|\mathbf{k}|_2^2/2).$$

□

Example 4.2 Periodization of the Hardy multiquadric.[1]

$$G(\mathbf{x}) = \sum_{\mathbf{k} \in \mathbb{Z}^q} (\alpha^2 + |\mathbf{x} - 2\pi \mathbf{k}|_2^2)^{-1}, \quad \hat{G}(\mathbf{k}) = \frac{\pi^{(q+1)/2}}{\Gamma\left(\frac{q+1}{2}\right)\alpha} \exp(-\alpha|\mathbf{k}|_2).$$

□

---

[1] A Hardy multiquadric is a function of the form $\mathbf{x} \to (\alpha^2 + |\mathbf{x}|_2^2)^{-1}$, $\mathbf{x} \in \mathbb{R}^q$. It is one of the oft-used function in theory and applications of radial basis function networks. For a survey, see the paper [27] of Hardy.

Example 4.3 Tensor product construction using the Poisson kernel. With $0 < r < 1$,

$$G(\mathbf{x}) = \prod_{j=1}^{q} \frac{1 - r^2}{1 + r^2 - 2r\cos(x_j)}, \quad \hat{G}(\mathbf{k}) = r^{|\mathbf{k}|_1}.$$

$\square$

For the proof of Theorem 4.1, we recall the following Nikolskii inequalities, see, e.g., [77, Section 4.9.4, p. 231].

Lemma 4.1 *Let* $1 \le p < r \le \infty$. *Then for* $n \in \mathbb{N}$,

$$\|P\|_r \le cn^{\left(\frac{q}{p} - \frac{q}{r}\right)} \|P\|_p, \quad P \in \mathbb{H}_n^q. \tag{4.7}$$

We are now ready to prove Theorem 4.1. In the remainder of this section, we fix an infinitely differentiable low pass filter h.

Proof of Theorem 4.1 In this proof, let $R = \sigma_N(h, G)$, so that $R \in \mathbb{H}_N^q$, and

$$\|G - R\|_\infty \le cE_{N,\infty}(G). \tag{4.8}$$

Since ν is a quadrature measure of order $n + N$ and $D_G(P) \in \mathbb{H}_n^q$, we obtain

$$P(\mathbf{x}) = \int_{\mathbb{T}^q} G(\mathbf{x} - \mathbf{y})D_G(P, \mathbf{y})\,d\mu_q^*(\mathbf{y})$$

$$= \int_{\mathbb{T}^q} R(\mathbf{x} - \mathbf{y})D_G(P, \mathbf{y})\,d\mu_q^*(\mathbf{y}) + \int_{\mathbb{T}^q} \left[G(\mathbf{x}-\mathbf{y}) - R(\mathbf{x}-\mathbf{y})\right]D_G(P, \mathbf{y})\,d\mu_q^*(\mathbf{y})$$

$$= \int_{\mathbb{T}^q} R(\mathbf{x} - \mathbf{y})D_G(P, \mathbf{y})\,d\nu(\mathbf{y}) + \int_{\mathbb{T}^q} \left[G(\mathbf{x} - \mathbf{y}) - R(\mathbf{x} - \mathbf{y})\right]D_G(P, \mathbf{y})\,d\mu_q^*(\mathbf{y})$$

$$= \int_{\mathbb{T}^q} G(\mathbf{x} - \mathbf{y})D_G(P, \mathbf{y})\,d\nu(\mathbf{y}) + \int_{\mathbb{T}^q} \left[G(\mathbf{x} - \mathbf{y}) - R(\mathbf{x} - \mathbf{y})\right]D_G(P, \mathbf{y})\,d\mu_q^*(\mathbf{y})$$

$$- \int_{\mathbb{T}^q} \left[G(\mathbf{x} - \mathbf{y}) - R(\mathbf{x} - \mathbf{y})\right]D_G(P, \mathbf{y})\,d\nu(\mathbf{y}).$$

Consequently,

$$\left|P(\mathbf{x}) - \int_{\mathbb{T}^q} G(\mathbf{x} - \mathbf{y})D_G(P, \mathbf{y})\,d\nu(\mathbf{y})\right|$$

$$\le \int_{\mathbb{T}^q} |G(\mathbf{x} - \mathbf{y}) - R(\mathbf{x} - \mathbf{y})|\,|D_G(P, \mathbf{y})|\,d\mu_q^*(\mathbf{y})$$

$$+ \int_{\mathbb{T}^q} |G(\mathbf{x} - \mathbf{y}) - R(\mathbf{x} - \mathbf{y})|\,|D_G(P, \mathbf{y})|\,d|\nu|(\mathbf{y}).$$

Since v is an M–Z quadrature measure of order $n + N$, $\|v\|_{n+N} \geq 1$, and we conclude that

$$\left\| P - \int_{\mathbb{T}^q} G(\circ - \mathbf{y}) D_G(P, \mathbf{y}) \, dv(\mathbf{y}) \right\|_\infty \leq c E_{N,\infty}(G) \left\{ \|D_G(P)\|_1 + \|D_G(P)\|_{v;1} \right\}$$

$$\leq c \|v\|_{n+N} E_{N,\infty}(G) \|D_G(P)\|_1. \qquad (4.9)$$

Using Nikolskii inequalities (4.7) and recalling the definition (4.1) of $\widehat{D_G(P)}(\mathbf{k})$, we deduce that

$$\|D_G(P)\|_1^2 \leq \|D_G(P)\|_2^2 = \sum_{\mathbf{k}:|\mathbf{k}|_2 < n} \left| \frac{\hat{P}(\mathbf{k})}{\hat{G}(\mathbf{k})} \right|^2 \leq \mathfrak{m}_n^{-2} \sum_{\mathbf{k}:|\mathbf{k}|_2 < n} |\hat{P}(\mathbf{k})|^2 = \mathfrak{m}_n^{-2} \|P\|_2^2$$

$$\leq c \left(n^{2\left(\frac{q}{p} - \frac{q}{2}\right)_+} \right) \mathfrak{m}_n^{-2} \|P\|_p^2.$$

The estimate (4.4) follows by substituting the above estimate into (4.9). $\qquad \square$

While Theorem 4.1 shows that polynomials can be approximated well by PBF networks, the converse is also true. To describe this, we introduce some notation.

Let $\{\mathbf{y}_j\}_{j=1}^M \subset \mathbb{T}^q$ and let $n \in \mathbb{N}$ be an integer with

$$\min_{j \neq k} |\mathbf{y}_j - \mathbf{y}_k| \geq 1/n. \qquad (4.10)$$

We note that this implies $M \leq cn^q$. In the sequel, we will assume tacitly that $\{\mathbf{y}_j\}_{j=1}^M$ is one of the members of a sequence of finite subsets of $\mathbb{T}^q$. We assume that M and n are variables, and then the following constants are independent of these.

Theorem 4.2 *Let $\{\mathbf{y}_j\}_{j=1}^M \subset \mathbb{T}^q$, $n \in \mathbb{N}$ be an integer satisfying (4.10). Let $r \geq 0$, $\mathbf{a} \in \mathbb{R}^M$, and let*

$$\mathcal{G}(\mathbf{x}) = \sum_{j=1}^M a_j G(\mathbf{x} - \mathbf{y}_j), \qquad \mathbf{x} \in \mathbb{T}^q.$$

Then for $N \in \mathbb{N}$, we have

$$\|(-\Delta)^{r/2}\mathcal{G} - (-\Delta)^{r/2}\sigma_N(h, \mathcal{G})\|_\infty \leq c_1 \left(n^{\left(\frac{q}{p} - q\right)_+} \right) \frac{E_{N,\infty}((-\Delta)^{r/2}G)}{\mathfrak{m}_{cn}((-\Delta)^{r/2}G)} \|(-\Delta)^{r/2}\mathcal{G}\|_p.$$

$$(4.11)$$

In particular, if $G \in \mathcal{E}_A$ for some $A > 0$, then there exists a $\rho = \rho(A, p, r, G) \in (0, 1)$ such that

$$\|(-\Delta)^{r/2}\mathcal{G} - (-\Delta)^{r/2}\sigma_{cAn}(h, \mathcal{G})\|_\infty \leq c_1 \rho^n \|(-\Delta)^{r/2}\mathcal{G}\|_p. \qquad (4.12)$$

Before proceeding to the proof, we note the following corollary.

Corollary 4.3 *Let $A > 0$ and $G \in \mathcal{E}_A$. With notation as in Theorem 4.2, we have for $n \in \mathbb{N}$,*

$$\|(-\Delta)^{r/2}\mathcal{G}\|_p \le cn^r\|\mathcal{G}\|_p, \tag{4.13}$$

where c is a positive constant, independent of the choice of the coefficients or the nodes $\{\mathbf{y}_j\}$, as long as (4.10) holds.

Proof If n is large enough, (4.12) yields

$$\|(-\Delta)^{r/2}\mathcal{G} - (-\Delta)^{r/2}\sigma_{cAn}(h, \mathcal{G})\|_p \le (1/2)\|(-\Delta)^{r/2}\mathcal{G}\|_p, \tag{4.14}$$

and since we may also use the same estimate with $r = 0$,

$$\|\mathcal{G} - \sigma_{cAn}(h, \mathcal{G})\|_p \le (1/2)\|\mathcal{G}\|_p. \tag{4.15}$$

This leads to

$$\|(-\Delta)^{r/2}\mathcal{G}\|_p \le 2\|(-\Delta)^{r/2}\sigma_{cAn}(h, \mathcal{G})\|_p, \quad \|\sigma_{cAn}(h, \mathcal{G})\|_p \le (3/2)\|\mathcal{G}\|_p. \tag{4.16}$$

Then using the Bernstein-type inequality in Theorem 3.2(b), we obtain

$$\|(-\Delta)^{r/2}\mathcal{G}\|_p \le 2\|(-\Delta)^{r/2}\sigma_{cAn}(h, \mathcal{G})\|_p \le c_1 n^r \|\sigma_{cAn}(h, \mathcal{G})\|_p \le c_2 n^r \|\mathcal{G}\|_p.$$

If n is not large enough, then (4.13) is obtained by adjusting the constant factor.

The proof of Theorem 4.2 is yet another interesting application of the localization estimate (3.2). Using this estimate in a very critical manner, we proved the following in [5, Theorem 6.2].

Proposition 4.2 *Let $\{\mathbf{y}_j\}_{j=1}^M \subset \mathbb{T}^q$ and let $n \in \mathbb{N}$ be an integer satisfying (4.10). Let $1 \le p \le \infty$ and $\mathbf{a} \in \mathbb{R}^M$. Then we have*

$$c_2 n^{q/p'}|\mathbf{a}|_p \le \left\|\sum_{j=1}^M a_j \Phi_m(h, \circ - \mathbf{y}_j)\right\|_p \le c_3 n^{q/p'}|\mathbf{a}|_p, \tag{4.17}$$

for $m \ge c_1 n$.

Proof of Theorem 4.2 We observe first that is enough to prove the theorem for $r = 0$. The general case follows by applying the result with the activation function $(-\Delta)^{r/2}G$ in place of G. Let $\mathbf{x} \in \mathbb{T}^q$. We note that

$$\sigma_N(h, \mathcal{G}, \mathbf{x} - \mathbf{y}_j) = \sum_{k=1}^M a_j \sigma_N(h, G, \mathbf{x} - \mathbf{y}_j),$$

and $M \leq cn^q$, so that

$$\|\mathcal{G} - \sigma_N(h, \mathcal{G})\|_\infty \leq cE_{N,\infty}(G)|\mathbf{a}|_1 \leq cn^{q/2}E_{N,\infty}(G)|\mathbf{a}|_2. \qquad (4.18)$$

To estimate $|\mathbf{a}|_2$, we fix $m = Cn$ for a sufficiently large m so as to satisfy the condition in Proposition 4.2. Let α be defined by

$$\alpha \overset{\text{def}}{=} \left(\frac{q}{p} - \frac{q}{2} \right)_+.$$

Then we can deduce with some calculation and with (4.17) applied with $p = 2$ that

$$\|\mathcal{G}\|_p^2 \geq c\|\sigma_m(h, \mathcal{G})\|_p^2 \geq cn^{-2\alpha}\|\sigma_m(h, \mathcal{G})\|_2^2$$

$$= cn^{-2\alpha} \sum_{j,\ell=1}^{M} a_j \overline{a_\ell} \sum_{\mathbf{k}\in\mathbb{Z}^q} h\left(\frac{|\mathbf{k}|_2}{m}\right)^2 |\hat{G}(\mathbf{k})|^2 \exp(-i\mathbf{k} \cdot (\mathbf{y}_j - \mathbf{y}_\ell))$$

$$= cn^{-2\alpha} \sum_{\mathbf{k}\in\mathbb{Z}^q} h\left(\frac{|\mathbf{k}|_2}{Cn}\right)^2 |\hat{G}(\mathbf{k})|^2 \left| \sum_{j=1}^{M} a_j \exp(-i\mathbf{k} \cdot \mathbf{y}_j) \right|^2$$

$$\geq cn^{-2\alpha} \mathfrak{m}_{Cn}(G)^2 \sum_{\mathbf{k}\in\mathbb{Z}^q} h\left(\frac{|\mathbf{k}|_2}{m}\right)^2 \left| \sum_{j=1}^{M} a_j \exp(-i\mathbf{k} \cdot \mathbf{y}_j) \right|^2$$

$$= cn^{-2\alpha} \mathfrak{m}_{Cn}(G)^2 \left\| \sum_{j=1}^{M} a_j \Phi_m(h, \circ - \mathbf{y}_j) \right\|_2^2$$

$$\geq cn^{-2\alpha+q} \mathfrak{m}_{Cn}(G)^2 |\mathbf{a}|_2^2.$$

Thus,

$$n^{q/2}|\mathbf{a}|_2 \leq cn^\alpha \mathfrak{m}_{Cn}(G)^{-1} \|\mathcal{G}\|_p.$$

Finally, the estimate (4.11) is obtained by substituting the upper bound on $|\mathbf{a}|_2$ from this estimate into (4.18). □

4.2 Direct and equivalence theorems

In this and the next subsections, we fix $G \in \mathcal{E}_A$. We will deal with two different sequences of measures; the sequence ν will be a sequence of discrete measures, whose supports will give us the centers of the networks, and the sequence μ will be the sequence of measures so that the information on the target function is given in terms of integrals with respect to the members of this sequence. It is a common practice in the literature on RBF networks to choose the centers the same as the points at which the target function is sampled, but we wish to make it a point that this is not necessary. To avoid confusion, we make the following definition.

Definition 4.2 Let $A > 0$, and $\boldsymbol{\nu} = \{\nu_j\}_{j=0}^{\infty}$ be a sequence of Borel measures. We say that $\boldsymbol{\nu} \in \mathfrak{M}_A$ if each of the following conditions is satisfied. Here, the constants may depend upon A and the sequence $\boldsymbol{\nu}$, but not on j or the individual measures ν_j.

1. Each ν_j is a discrete measure, and the support $\mathcal{C}_j = \mathsf{supp}(\nu_j)$ satisfies

$$\delta(\mathcal{C}_j) \sim \eta(\mathcal{C}_j) \sim 2^{-j}.$$

2. Each ν_j is an M–Z quadrature measure of order $(1 + A)2^j$, and $\|\nu\|_{(1+A)2^j} \le c$.
3. If $j > m$ then $\mathcal{C}_m \subseteq \mathcal{C}_j$.

We point out an important example of measures in $\mathfrak{M}_A$. Let

$$ZD_j^q = \{\mathbf{k} \in \mathbb{Z}^q : -2^j + 1 \le k_\ell \le 2^j, \ \ell = 1, 2, \dots, q\}, \quad j = 0, 1, 2, \dots,$$

and

$$\mathbf{v}_{\mathbf{k},j} = \mathbf{k}\pi/2^j, \quad \mathbf{k} \in ZD_j^q, \ j = 0, 1, 2, \dots. \tag{4.19}$$

It can be verified using the corresponding univariate result repeatedly that the following quadrature formula holds.

$$\frac{1}{(2\pi)^q} \int_{\mathbb{T}^q} P(\mathbf{t})\, d\mathbf{t} = \frac{1}{2^{q(j+1)}} \sum_{\mathbf{k} \in ZD_j^q} P(\mathbf{v}_{\mathbf{k},j}), \quad P \in \mathbb{H}_{2^{j+1}}^q. \tag{4.20}$$

Let m be the integer part of $\log_2(1 + A)$, and for $j = 0, 1, \dots$, let ν_j^* be the measure ν_j^* that associates the mass $2^{-q(j+m+1)}$ with each $\mathbf{v}_{\mathbf{k},j+m}$, $\mathbf{k} \in ZD_{j+m}^q$. Then the sequence $\boldsymbol{\nu}^* = \{\nu_j^*\}_{j=0}^{\infty} \in \mathfrak{M}_A$. This fact can be checked using Lemma 2.1 for each coordinate.

Let $\boldsymbol{\nu} \in \mathfrak{M}_A$. We define

$$\mathbb{G}_j = \mathbb{G}_j(\boldsymbol{\nu}) \overset{\text{def}}{=} \{G(\circ - \mathbf{y}) : \mathbf{y} \in \mathsf{supp}(\nu_j)\}, \quad j \in \mathbb{N}, \quad \mathbb{G}_0 = \{0\}, \tag{4.21}$$

and observe that, since $\{\mathsf{supp}(\nu_j)\}$ is a nested sequence, $\mathbb{G}_j$ is a nested sequence of linear subspaces of X^{∞} and that the closure of their union in the sense of L^p is X^p. For $f \in L^p$, we write

$$\text{dist}\,(L^p, f, \mathbb{G}_j) = \inf\{\|f - \mathcal{G}\|_p : \mathcal{G} \in \mathbb{G}_j\}.$$

In addition to the centers, we will assume in the remainder of this and the next subsection that $\boldsymbol{\mu} = \{\mu_j\}_{j=0}^{\infty}$ is a sequence of measures such that each μ_j is an M–Z quadrature measure of order $3 \times 2^{j-1}$. The sequence $\boldsymbol{\mu}^*$ in which each $\mu_j = \mu_q^*$ is one such sequence. Our interest is both in this sequence and the case when the μ_j's are discretely supported.

The direct theorem can be stated as follows.

Theorem 4.3 *Let $A > 0$, $G \in \mathcal{E}_A$, $\boldsymbol{v} \in \mathfrak{M}_A$. Then there exists $\rho \in (0, 1)$ such that for $1 \le p \le \infty$ and $f \in L^p$ we have*

$$\operatorname{dist}(L^p, f, \mathbb{G}_j(\boldsymbol{v})) \le \left\| f - \int_{\mathbb{T}^q} G(\circ - \mathbf{y}) D_G\left(\sigma_{2^j}(\mu_q^*; h, f), \mathbf{y}\right) d\nu_j(\mathbf{y}) \right\|_p$$

$$\le c\{E_{2^j,p}(f) + \rho^{2^j}\|f\|_p\}. \tag{4.22}$$

Consequently, if $\gamma > 0$ and $f \in W_{\gamma,p}$, then

$$\operatorname{dist}(L^p, f, \mathbb{G}_j(\boldsymbol{v})) = \mathcal{O}(2^{-j\gamma}). \tag{4.23}$$

In the case when $p = \infty$, (4.22) remains valid provided that $\sigma_{2^j}(\mu_q^*; h, f)$ is replaced by $\sigma_{2^j}(\mu_j; h, f)$, where $\boldsymbol{\mu}$ is a sequence as described earlier.

We note the following variations of (4.22).

Corollary 4.4 *With the set up as in Theorem 4.3,*

$$\left\| f - \int_{\mathbb{T}^q} G(\circ - \mathbf{y}) D_G\left(\sigma_{2^j}(\mu_q^*; h, f), \mathbf{y}\right) d\nu_j^*(\mathbf{y}) \right\|_p \le c\{E_{2^j,p}(f) + \rho^{2^j}\|f\|_p\}. \tag{4.24}$$

In the case when $p = \infty$ and $\boldsymbol{\mu} \in \mathfrak{M}_{\max(A,4)}$, we also have

$$\left\| f - \int_{\mathbb{T}^q} G(\circ - \mathbf{y}) D_G\left(\sigma_{2^j}(\mu_j; h, f), \mathbf{y}\right) d\mu_j(\mathbf{y}) \right\|_\infty \le c\{E_{2^j,\infty}(f) + \rho^{2^j}\|f\|_\infty\}. \tag{4.25}$$

We observe that the integral expression in (4.24) is a PBF network with centers at the dyadic points (4.19). This is in keeping with traditional results on PBF approximation where scaled integer translates are considered. However, unlike in the classical setting, we do not require any a priori conditions such as the Strang–Fix conditions to ensure any polynomial reproduction. In particular, constants are not included in the network. In the case when μ_j is used in place of μ_q^*, the information on f which is used in the construction of the network is the values of f at points in $\operatorname{supp}(\mu_j)$, but we have different choices for the centers of the resulting network. A popular choice is to use points from the same sequence of data sets as centers as in (4.25). However, we may also use dyadic centers instead as in (4.24), potentially leading to faster computations.

All the constructions are linear operators on the information available on the target function. Finally, the bounds (4.22), (4.24), and (4.25), and also their variants where $\sigma_{2^j}(\mu_q^*; h, f)$ is replaced by $\sigma_{2^j}(\mu_j; h, f)$, show that the PBF networks constructed there are bounded operators on the spaces involved. Therefore, the constructions are stable.

Proof of Theorem 4.3 We need only to sketch the proof of (4.22); the other statements are immediate consequences. To prove (4.22), we use (4.6) with $\sigma_{2^j}(\mu_q^*; h, f)$ (or with $\sigma_{2^j}(\mu_j; h, f)$ respectively, when appropriate), recall that $\|v_j\|_{2^j} \sim \|v_j\|_{(1+A)2^j} \leq c$, and then use the resulting estimate together with the triangle inequality and (3.4) (or with (3.38) respectively) to arrive at (4.22). □

Converse theorems for approximation by elements of $\mathbb{G}_j$ can be obtained using Corollary 4.3, cf. [12, Theorem 9.1, also Chapter 6.7]. We note only the analogue of the equivalence theorem without proof.

Theorem 4.4 *Let* $1 \leq p \leq \infty$, $\gamma > 0$, $A > 0$, $G \in \mathcal{E}_A$ *and* $v \in \mathfrak{M}_A$. *Then the following are equivalent.*

(a) $f \in W_{\gamma, p}$.
(b) $\mathrm{dist}\,(L^p, f, \mathbb{G}_j(v)) = \mathcal{O}(2^{-j\gamma})$.
(c) *We have*

$$\left\| f - \int_{\mathbb{T}^q} G(\circ - \mathbf{y}) D_G\left(\sigma_{2^j}(\mu_q^*; h, f), \mathbf{y}\right) dv_j^*(\mathbf{y}) \right\|_p = \mathcal{O}(2^{-j\gamma}). \quad (4.26)$$

(d) *If* $p = \infty$, *then each of the above statements is also equivalent to*

$$\left\| f - \int_{\mathbb{T}^q} G(\circ - \mathbf{y}) D_G\left(\sigma_{2^j}(\mu_j; h, f), \mathbf{y}\right) dv_j(\mathbf{y}) \right\|_\infty = \mathcal{O}(2^{-j\gamma}).$$

We remark that the statement (a) in the above theorem is independent of the sequence v. Therefore, the part (a) above implies part (b) for *every* $v \in \mathfrak{M}_A$. On the other hand, if part (b) holds for *some* $v \in \mathfrak{M}_A$, then part (a) holds. In particular, when approximating a function from $W_{\gamma, p}$, the asymptotic degree of approximation by PBF networks does not depend upon the choice of centers and weights—encoded in the measures v—as long as $v \in \mathfrak{M}_A$.

4.3 Wavelet-like representation using PBFs

The wavelet-like representations using PBF networks with activation function in $\mathcal{E}_A$ for some $A > 0$ are very similar to those for multivariate trigonometric polynomials. This can be shown by using Theorem 4.1. We summarize them below. We will sketch the proofs only, except for the proof of the frame property which requires some new ideas.

In the sequel, let $A > 0$ and let $G \in \mathcal{E}_A$ be fixed. Also, let h be a fixed, infinitely differentiable low pass filter. We will also make the same assumptions for the sequence of measures $\boldsymbol{\nu} = \{\nu_j\}_{j=0}^{\infty}$ and $\boldsymbol{\mu} = \{\mu_j\}_{j=0}^{\infty}$; namely, $\boldsymbol{\nu} \in \mathfrak{M}_A$ and each μ_j is an M–Z quadrature measure of order $3 \times 2^{j-1}$.

First, we consider the operators

$$\mathcal{S}_j^*(f, \mathbf{x}) \overset{\text{def}}{=} \int_{\mathbb{T}^q} G(\mathbf{x} - \mathbf{y}) D_G \left(\sigma_{2^j}(\mu_q^*; h, f), \mathbf{y} \right) d\nu_j(\mathbf{y}), \quad j = 0, 1, \ldots, \ f \in L^p, \ p \in [1, \infty],$$

(4.27)

and the corresponding frame operators

$$T_j^*(f, \mathbf{x}) = \begin{cases} \mathcal{S}_0^*(f, \mathbf{x}), & \text{if } j = 0, \\ \mathcal{S}_j^*(f, \mathbf{x}) - \mathcal{S}_{j-1}^*(f, \mathbf{x}), & \text{if } j \in \mathbb{N}. \end{cases}$$

(4.28)

The variants when the information about f is in terms of the sequence $\boldsymbol{\mu}$ are given by

$$\mathcal{S}_j(f, \mathbf{x}) \overset{\text{def}}{=} \int_{\mathbb{T}^q} G(\mathbf{x} - \mathbf{y}) D_G \left(\sigma_{2^j}(\mu_j; h, f), \mathbf{y} \right) d\nu_j(\mathbf{y}), \quad j = 0, 1, \ldots, \ f \in X^{\infty},$$

(4.29)

and

$$T_j(f, \mathbf{x}) = \begin{cases} \mathcal{S}_0(f, \mathbf{x}), & \text{if } j = 0, \\ \mathcal{S}_j(f, \mathbf{x}) - \mathcal{S}_{j-1}(f, \mathbf{x}), & \text{if } j \in \mathbb{N}. \end{cases}$$

(4.30)

Theorem 4.1 leads immediately to the following estimates for these operators.

Proposition 4.3 *There exists $\rho \in (0, 1)$ such that, if $1 \le p \le \infty$ and $f \in X^p$, then we have*

$$\|\mathcal{S}_j^*(f) - \sigma_{2^j}(\mu_q^*; h, f)\|_p \le c\rho^{2^j} \|\sigma_{2^j}(\mu_q^*; h, f)\|_p \le c\rho^{2^j} \|f\|_p. \quad (4.31)$$

For any measure $\tilde{\nu}$ on $\mathbb{T}^q$,

$$\|T_j^*(f) - \tau_j(\mu_q^*; h, f)\|_{\tilde{\nu}; p} \le c\rho^{2^j} \|f\|_p. \quad (4.32)$$

Analogous statements hold if $p = \infty$ and $\mathcal{S}_j^*(f)$ (respectively, $T_j^*(f)$, $\sigma_{2^j}(\mu_q^*; h, f)$, $\tau_j(\mu_q^*; h, f)$) is replaced by $\mathcal{S}_j(f)$ (respectively, $T_j(f)$, $\sigma_{2^j}(\mu_j; h, f)$, $\tau_j(\boldsymbol{\mu}; h, f)$).

The following two theorems, analogous to Theorems 4.5 and 3.10, are immediate consequences of these theorems and the above proposition, except for part (b) of Theorem 4.5 below.

Theorem 4.5 *Let $1 \le p \le \infty$, $\gamma > 0$, $f \in X^p$, and let G and the measures $\boldsymbol{\nu}$, $\boldsymbol{\mu}$ be as described earlier.*

(a) *We have*

$$f = \sum_{j=0}^{\infty} T_j^*(f) \qquad (4.33)$$

with convergence in the sense of X^p.
(b) *If $f \in L^2$ then*

$$\|f\|_2^2 \sim \sum_{j=0}^{\infty} \|T_j^*(f)\|_2^2, \qquad (4.34)$$

where the constants are independent of f, but may depend upon G.
(c) *If $f \in X^{\infty}$, then we have the uniformly convergent expansion*

$$f = \sum_{j=0}^{\infty} T_j(f) \qquad (4.35)$$

The analogue of Theorem 3.10 is the following:

Theorem 4.6 *Let $1 \le p \le \infty$, $\gamma > 0$, $\mathbf{x}_0 \in \mathbb{T}^q$, $f \in X^p$, and let G and the measures v, μ be as described earlier. Then the following are equivalent.*

(a) *$f \in W_{\gamma,p}(\mathbf{x}_0)$.*

(b) *There exists an arc $I \ni \mathbf{x}_0$ such that*

$$\|T_j^*(f)\|_{v_j; p, I} = \mathcal{O}(2^{-j\gamma}). \qquad (4.36)$$

(c) *In the case when $p = \infty$, one may replace (4.36) by*

$$\max_{\mathbf{t} \in \text{supp}(v_j) \cap I} |T_j(f, \mathbf{t})| = \mathcal{O}(2^{-j\gamma}).$$

Part (b) of Theorem 4.5 is not immediately obvious. In order to prove this part, we first state a lemma.

Lemma 4.2 *Let $f \in L^2$, and let ρ be as in Proposition 4.3. Then for $N \in \mathbb{N}$ and $j \in \mathbb{N}$,*

$$(1/4)\|f\|_2^2 \le \|\sigma_{2^N}(\mu_q^*; h, f)\|_2^2 + \sum_{j=N+1}^{\infty} \|\tau_j(\mu_q^*; h, f)\|_2^2 \le \|f\|_2^2, \quad (4.37)$$

and

$$\left| \|T_j^*(f)\|_2^2 - \|\tau_j(\mu_q^*; h, f)\|_2^2 \right| \le c\rho^{2^j}\|f\|_2^2. \qquad (4.38)$$

Proof Using the same argument as in the proof of (2.36) in the proof of Theorem 2.5, we deduce that for all $\mathbf{k} \in \mathbb{Z}^q$

$$1 \leq 2 \left(h \left(\frac{|\mathbf{k}|_2}{N} \right) \right)^2 + 4 \sum_{j=N+1}^{\infty} \left(h \left(\frac{|\mathbf{k}|_2}{2^j} \right) - h \left(\frac{|\mathbf{k}|_2}{2^{j-1}} \right) \right)^2 \leq 4.$$

Then, again as in the proof of Theorem 2.5, we use the Parseval identity to obtain (4.37).

Next, we use (4.32) with $\nu = \mu_q^*$, $A = \mathbb{T}^q$, and $p = 2$ to observe that

$$\left| \|\mathcal{T}_j^*(f)\|_2^2 - \|\tau_j(\mu_q^*; h, f)\|_2^2 \right| = (\|\mathcal{T}_j^*(f)\|_2 + \|\tau_j(\mu_q^*; h, f)\|_2)(\|\mathcal{T}_j^*(f)\|_2 - \|\tau_j(\mu_q^*; h, f)\|_2)$$

$$\leq c\|f\|_2 \|\mathcal{T}_j^*(f) - \tau_j(\mu_q^*; h, f)\|_2 \leq c\rho^{2^j} \|f\|_2^2.$$

This is precisely (4.38).   □

Proof of Theorem 4.5 (b). Let $N \in \mathbb{N}$ be fixed but to be chosen later. In light of (4.38),

$$\left| \sum_{j=N+1}^{\infty} \|\mathcal{T}_j^*(f)\|_2^2 - \sum_{j=N+1}^{\infty} \|\tau_j(\mu_q^*; h, f)\|_2^2 \right| \leq c\|f\|_2^2 \sum_{j=N+1}^{\infty} \rho^{2^j}. \quad (4.39)$$

Also, (4.31) implies that

$$\left| \|\mathcal{S}_N^*(f)\|_2 - \|\sigma_{2^N}(\mu_q^*; h, f)\|_2 \right| \leq c_1 \rho^{2^N} \|\sigma_{2^N}(\mu_q^*; h, f)\|_2. \quad (4.40)$$

We now choose N such that

$$c \sum_{j=N+1}^{\infty} \rho^{2^j} < 1/8, \quad c_1 \rho^{2^N} < 1/2, \quad (4.41)$$

where c and c_1 are as in the previous two displayed estimates. Then

$$\|\mathcal{S}_N^*(f)\|_2 \leq (3/2)\|\sigma_{2^N}(\mu_q^*; h, f)\|_2 \leq 3\|\mathcal{S}_N^*(f)\|_2. \quad (4.42)$$

With this preparation, we first prove the upper bound on $\|f\|_2^2$. The first inequality in (4.37) yields

$$(1/4)\|f\|_2^2 \leq 4\|\mathcal{S}_N^*(f)\|_2^2 + \sum_{j=N+1}^{\infty} \|\mathcal{T}_j^*(f)\|_2^2 + (1/8)\|f\|_2^2,$$

that is,

$$(1/32)\|f\|_2^2 \leq \|\mathcal{S}_N^*(f)\|_2^2 + \sum_{j=N+1}^{\infty} \|\mathcal{T}_j^*(f)\|_2^2. \quad (4.43)$$

Since

$$\|\mathcal{S}_N^*(f)\|_2^2 \le \left(\sum_{j=0}^N \|\mathcal{T}_j^*(f)\|_2 \right)^2 \le (N+1) \sum_{j=0}^N \|\mathcal{T}_j^*(f)\|_2^2,$$

(4.43) shows that

$$c\|f\|_2^2 \le \sum_{j=0}^\infty \|\mathcal{T}_j^*(f)\|_2^2. \qquad (4.44)$$

The lower bound is easier. For each $j \ge 0$, we have $\|\mathcal{T}_j^*(f)\|_2 \le c\|f\|_2$. Therefore, using (4.37) and (4.39) and keeping in mind our choice of N as in (4.41), we can conclude that

$$\begin{aligned}
\sum_{j=0}^\infty \|\mathcal{T}_j^*(f)\|_2^2 &= \sum_{j=0}^N \|\mathcal{T}_j^*(f)\|_2^2 + \sum_{j=N+1}^\infty \|\mathcal{T}_j^*(f)\|_2^2 \\
&\le c(N+1)\|f\|_2^2 + \sum_{j=N+1}^\infty \|\tau_j(\mu_q^*; h, f)\|_2^2 + (1/8)\|f\|_2^2 \\
&\le (c(N+1) + 1 + 1/8)\|f\|_2^2.
\end{aligned}$$

This completes the proof.   □

We note in closing that with a proper normalization, we may assume that $\hat{G}(0) = 1$. Then $\mathfrak{m}_n(G) \ge 1$ for all n. Hence, the statement that $G \in \mathcal{E}_A$ implies, in particular, that

$$\limsup_{m \to \infty} E_{m,\infty}(G)^{1/m} < 1.$$

This, in turn, implies that G is analytic on $\mathbb{T}^q$. There are many examples of activation functions which are used in practice, notably the periodization of the Wendland functions, or Green's functions of the operators $(-\Delta)^{r/2}$, for which this condition is not satisfied. The analogues of Theorems 4.1 and 4.2 are then much weaker; they are given in [49] in a very general context. Direct and converse theorems for approximation with networks with such activation functions are also obtained in that paper. However, the estimates there are not strong enough to obtain the characterization of local smoothness classes.

It is worthwhile to comment about the relationship of our results with those in the papers [14] by Dũng and Micchelli and [40] of Maiorov.

- The paper [40] deals with approximation in L^2, and the paper [14] deals with L^p, $1 < p < \infty$. Our paper includes both $p = 1$ and the case of continuous functions when $p = \infty$.

- The paper [14] deals with Korobov spaces rather than Sobolev spaces in the sense of our definitions. While our methods can be extended to the case of Korobov spaces, the papers in their present form are not comparable.
- The papers [14,40] give the bounds on approximation in terms of the number of neurons. Our paper gives the bounds in terms of the minimal separation among centers. In the case of uniform grids, the two concepts coincide. In this case, the upper bound in [40] is similar to ours, and is the ideas behind its proof are essentially similar to those used in this paper. Both of these ideas are originally developed in [51].
- The papers [14,40] give lower bounds in the sense of *worst case complexity*; i.e., lower distance bounds. Our focus is on individual functions. For example, the lower bound in Theorem 1.1 of [40] implies that there **exists some** f in the function class under consideration for which the lower bound applies. The converse theorems in this paper are conceptually quite different. They state that for each **individual function** f, without any prior knowledge of the class to which it belongs, the rate of decrease of the degree of approximation *implies* the smoothness class to which it belongs.
- The paper [14] depends upon dyadic decompositions as is customary in the study of Korobov spaces. Preliminary numerical experiments suggest that the hyperbolic cross versions of the operators considered in [14] (or our operators for that matter) are not localized. We use the term wavelet-like representation to mean that the coefficients characterize local smoothness classes analogous to classical wavelet expansions. In this sense, it is an open problem to obtain a wavelet-like representation that characterizes local Korobov spaces.
- There is an impressive lower bound in [14] which suggests that the best activation function in the case of approximation of Korobov spaces is the Korobov kernel itself. Intuitively, this deep result is somewhat expected, since the Korobov spaces are, in a limiting sense, translation networks with the Korobov kernel as the activation function. Analogous results for approximation of periodic functions are given in [52,53]. The current paper does not deal with the question of the choice of an optimal activation function.

5 Further extensions

5.1 Jacobi expansions

Given $\alpha > -1$ and $\beta > -1$, the Jacobi weights are defined by

$$w_{\alpha,\beta}(x) \overset{\text{def}}{=} \begin{cases} (1-x)^\alpha (1+x)^\beta, & \text{if } -1 < x < 1, \\ 0, & \text{otherwise.} \end{cases}$$

For $1 \le p < \infty$, the space $L^p(\alpha, \beta)$ is defined as the space of (equivalence classes of) functions f with

$$\|f\|_{\alpha,\beta;p} \overset{\text{def}}{=} \left(\int\limits_{-1}^{1} |f(x)|^p w_{\alpha,\beta}(x)\, dx \right)^{1/p} < \infty.$$

The symbol $X^p(\alpha,\beta)$ denotes $L^p(\alpha,\beta)$, if $1 \le p < \infty$, and $C([-1,1])$, the space of continuous functions on $[-1,1]$ with the maximum norm $\| \circ \|_\infty$, if $p = \infty$. The space of all algebraic polynomials of degree at most n will be denoted by Π_n.

There exists a unique system of orthonormalized Jacobi polynomials $\{ p_j^{(\alpha,\beta)}(x) = \gamma_j(\alpha,\beta)x^j + \cdots \}$, $\gamma_j(\alpha,\beta) > 0$ such that for integer $j, \ell = 0, 1, \ldots,$

$$\int\limits_{-1}^{1} p_j^{(\alpha,\beta)}(x) p_\ell^{(\alpha,\beta)}(x) w_{\alpha,\beta}(x)\, dx = \begin{cases} 1, & \text{if } j = \ell, \\ 0, & \text{otherwise.} \end{cases}$$

The uniqueness of the system implies that $p_j^{(\beta,\alpha)}(x) = (-1)^j p_j^{(\alpha,\beta)}(-x)$, $x \in \mathbb{R}$, $j = 0, 1, \ldots$. Therefore, we may assume in the sequel that $\alpha \ge \beta$. We will assume also that $\alpha \ge \beta \ge -1/2$.

If $f \in L^1(\alpha,\beta)$, then, in this subsection, we define the Jacobi coefficients by

$$\hat{f}(j) \overset{\text{def}}{=} \hat{f}(\alpha,\beta;j) \overset{\text{def}}{=} \int\limits_{-1}^{1} f(y) p_j^{(\alpha,\beta)}(y) w_{\alpha,\beta}(y)\, dy, \quad j = 0, 1, \ldots,$$

so that the formal Jacobi expansion of f is given by $\sum_{j \ge 0} \hat{f}(j) p_j^{(\alpha,\beta)}$. We define the summability operator analogous to σ_n as follows. Let $h : [0,\infty) \to \mathbb{R}$ be a compactly supported function. We define

$$\Phi_n(\alpha,\beta;h,x,y) = \sum_{j=0}^{\infty} h\left(\frac{j}{n}\right) p_j^{(\alpha,\beta)}(x) p_j^{(\alpha,\beta)}(y),$$

and, for $f \in L^1(\alpha,\beta)$,

$$\sigma_n(\alpha,\beta;h,f,x) = \sum_{j=0}^{\infty} h\left(\frac{j}{n}\right) \hat{f}(j) p_j^{(\alpha,\beta)}(x)$$

$$= \int\limits_{-1}^{1} f(y) \Phi_n(\alpha,\beta;h,x,y) w_{\alpha,\beta}(y)\, dy, \quad n \in \mathbb{N}.$$

Using a result on the Cesáro means of the Jacobi expansion, see below, it is fairly easy to show that these operators are uniformly bounded. To describe this, we first recall the definition of Cesáro means. If $\kappa > -1$, the Cesàro means of order κ of

$f \in L^1(\alpha, \beta)$ are defined by

$$C_n^{[\kappa]}(\alpha, \beta; f, x) \overset{\text{def}}{=} \binom{n + \kappa}{\kappa}^{-1} \sum_{j=0}^{n} \binom{n - j + \kappa}{\kappa} \hat{f}(j) p_j^{(\alpha, \beta)}(x) \tag{5.1}$$

The following theorem is well known, see, e.g., [1,76].

Theorem 5.1 *Let $\alpha, \beta \geq -1/2$, $\kappa > \max(\alpha, \beta) + 1/2$ be an integer, $1 \leq p \leq \infty$, and $f \in X^p(\alpha, \beta)$. Then*

$$\|C_n^{[\kappa]}(\alpha, \beta; f)\|_{\alpha, \beta; p} \leq c\|f\|_{\alpha, \beta; p}, \tag{5.2}$$

for $n \in \mathbb{N}$.

Using a summation by parts argument as done in [26, Theorem 71, p. 128], Theorem 5.1 (used with S in place of κ) leads immediately to the following corollary.

Corollary 5.1 *Let $S > \max(\alpha, \beta) + 1/2$. If $\{h_j\}$ is a sequence of real numbers, so that $h_j \to 0$ as $j \to \infty$, and*

$$\sum_{j=0}^{\infty} (j + 1)^S |\Delta^{S+1} h_j| < \infty, \tag{5.3}$$

where Δ here is the forward difference operator, then for $1 \leq p \leq \infty$ and $f \in X^p(\alpha, \beta)$,

$$\left\| \sum_{j=0}^{\infty} h_j \hat{f}(j) p_j^{(\alpha, \beta)} \right\|_p = \left\| \sum_{j=0}^{\infty} \binom{j + S}{S} C_j^{[S]}(\alpha, \beta; f) \Delta^{S+1} h_j \right\|_p$$

$$\leq c \left(\sum_{j=0}^{\infty} (j + 1)^S |\Delta^{S+1} h_j| \right) \|f\|_{\alpha, \beta; p}. \tag{5.4}$$

In particular, if h is a compactly supported and $S + 1$ times continuously differentiable function, then

$$\|\sigma_n(\alpha, \beta; h, f)\|_p \leq c\|f\|_p. \tag{5.5}$$

Using the same methods as in the trigonometric case and the above results, one can easily obtain direct and converse theorems as well as the wavelet-like representation theorems for characterization of suitably defined global smoothness classes; these results are formulated in [59]. However, these bounds by themselves are not sufficient to obtain a characterization of local smoothness. We proved the following localization estimates on the kernels Φ_n, see [45,48].

Theorem 5.2 *Let* $\alpha, \beta \geq -1/2$, $S \in \mathbb{N}$, *and let* $h_j = 0$ *for all sufficiently large* j. *Then*

$$a \left| \sum_{k=0}^{\infty} h_j p_j{}^{(\alpha,\beta)}(\cos\theta) p_j{}^{(\alpha,\beta)}(\cos\varphi) \right|$$

$$\leq c \sum_{j=0}^{\infty} \min\left((j+1), \frac{1}{|\theta-\varphi|} \right)^{\max(\alpha,\beta)+S+1/2} \times \sum_{m=1}^{S} (j+1)^{\max(\alpha,\beta)+1/2-S+m} |\Delta^m h_j|,$$

(5.6)

for $\theta, \varphi \in [0, \pi]$. *In particular, if* $h : [0, \infty) \to [0, \infty)$ *is a compactly supported function that can be expressed as an* S-*times iterated integral of a function of bounded total variation* V, *and if* $h'(t) = 0$ *in a neighborhood of* 0, *then*

$$|\Phi_n(\alpha, \beta; h, \cos\theta, \cos\varphi)| \leq c \, n^{2\max(\alpha,\beta)+2} \, V \min\left(1, \frac{1}{(n|\theta-\varphi|)^{\max(\alpha,\beta)+S+1/2}} \right),$$

(5.7)

for $n \in \mathbb{N}$.

A wavelet-like representation with characterization of local smoothness classes is given in [48].

The subject of Marcinkiewicz–Zygmund inequalities in this context is very well studied. Perhaps, the most classical example is a simple consequence of the Gauss–Jacobi quadrature formula. For $n \in \mathbb{N}$, let $\{x_{k,n}\}_{k=1}^{n}$ be the zeros of $P_n^{(\alpha,\beta)}$, and let

$$\lambda_{k,n} \stackrel{\text{def}}{=} \left\{ \sum_{j=0}^{n-1} p_j^{(\alpha,\beta)}(x_{k,n})^2 \right\}^{-1}, \quad k = 1, 2, \ldots, n.$$

Then the Gauss–Jacobi quadrature formula implies that

$$\sum_{k=1}^{n} \lambda_{k,n} |P(x_{k,n})|^2 = \int_{-1}^{1} |P(y)|^2 w_{\alpha,\beta}(y) \, dy, \quad P \in \Pi_{n-1}.$$

(5.8)

The following analogue in the case of L^p norms was proved in [68, Theorem 25, p. 168].

Theorem 5.3 *Let* $1 \leq p < \infty$ *and let* $c_1 > 0$. *Then there exists a constant* c *depending only on* α, β, *and* c_1, *such that for all* $m \in \mathbb{N}$ *and* $n \in \mathbb{N}$ *with* $1 \leq m \leq c_1 n$,

$$\sum_{k=1}^{n} \lambda_{k,n} |P(x_{k,n})|^p \leq c \|P\|_{\alpha,\beta;p}, \quad P \in \Pi_m.$$

(5.9)

This theorem found deep applications in investigations related to weighted mean convergence of Lagrange interpolation [67,68]. A survey of many of the classical results in this direction and their applications can be found in the paper [38] by Lubinsky. In [48], we proved the existence of M–Z quadrature measures based on arbitrary set of points on $[-1, 1]$ subject to a density condition, and we gave applications to wavelet-like representations based on values of the function at these points.

Characteristically for polynomial approximation, one can also construct localized operators which yield approximation commensurate with analyticity of the target function on intervals, rather than the much weaker smoothness conditions studied in the previous section. If $1 \leq p \leq \infty, n \geq 0$, and $f \in L^p(\alpha, \beta)$, then we define the degree of best weighted approximation of f by polynomials of degree at most n by

$$E_{n,p}(\alpha, \beta; f) \stackrel{\text{def}}{=} \min_{P \in \Pi_n} \|f - P\|_{\alpha, \beta; p}.$$

For integer $n \geq 1$, let the numbers $H^*_{j,n}$, $j = 0, \ldots, 5n - 1$ be defined by

$$\left(\frac{1+x}{2}\right)^n \Phi_{4n}(\alpha, \beta; x, 1) = \sum_{j=0}^{5n-1} H^*_{j,n} p_j^{(\alpha, \beta)}(x) p_j^{(\alpha, \beta)}(1), \qquad (5.10)$$

and let

$$\sigma_n(\alpha, \beta; H^*, f, x) \stackrel{\text{def}}{=} \sum_{j=0}^{5n-1} H^*_{j,n} \hat{f}(j) p_j^{(\alpha, \beta)}(x), \quad x \in [-1, 1].$$

In [19, Theorem 3.3], we proved the following.

Theorem 5.4 (a) *Let* $1 \leq p \leq \infty$, $\alpha, \beta \geq -1/2$, *and let* $f \in L^p(\alpha, \beta)$. *Then, with* H^* *as defined in* (5.10), *we have* $\sigma_n(\alpha, \beta; H^*, P) = P$ *for* $P \in \Pi_n$, *and* $\sup_{n \in \mathbb{N}} \|\sigma_n(\alpha, \beta; H^*, f)\|_{\alpha, \beta; p} \leq c\|f\|_{\alpha, \beta; p}$. *In addition,*

$$E_{5n,p}(\alpha, \beta; f) \leq \|f - \sigma_n(\alpha, \beta; H^*, f)\|_{\alpha, \beta; p} \leq c_1 E_{n,p}(\alpha, \beta; f). \quad (5.11)$$

(b) *Let* $f \in C([-1, 1])$, $x_0 \in [-1, 1]$, *and let* f *have an analytic continuation to a complex neighborhood of* x_0, *given by* $\{z \in \mathbb{C} : |z - x_0| \leq d\}$ *for some* d *with* $0 < d \leq 2$. *Then*

$$|f(x) - \sigma_n(\alpha, \beta; H^*, f, x)| \leq c(f, x_0) \exp\left(-c_1(d)n \frac{d^2 \log(e/2)}{e^2 \log(e^2/d)}\right),$$
$$x \in [x_0 - d/e, x_0 + d/e] \cap [-1, 1], \qquad (5.12)$$

where $\log$ *is the natural logarithm, and* e *is the basis of this logarithm.*

5.2 Approximation on the sphere

Theorems about Jacobi expansions translate easily into analogous theorems for approximation on the sphere. The following paragraph is taken from [61]. Let $q \in \mathbb{N}$, and let

$$\mathbb{S}^q \stackrel{\text{def}}{=} \left\{ (x_1, \ldots, x_{q+1}) \in \mathbb{R}^{q+1} : \sum_{j=1}^{q+1} x_j^2 = 1 \right\}.$$

A spherical cap, centered at $\mathbf{x}_0 \in \mathbb{S}^q$ and with radius α is defined by

$$\mathbb{S}_\alpha^q(\mathbf{x}_0) \stackrel{\text{def}}{=} \{ \mathbf{x} \in \mathbb{S}^q : \mathbf{x} \cdot \mathbf{x}_0 \geq \cos \alpha \}.$$

We note that for all $\mathbf{x}_0 \in \mathbb{S}^q$ we have $\mathbb{S}_\pi^q(\mathbf{x}_0) = \mathbb{S}^q$. In this subsection, the surface area (aka volume element) measure on $\mathbb{S}^q$ will be denoted by μ_q^* (there being no chance of confusion with the notation on the torus), and we write $\omega_q \stackrel{\text{def}}{=} \mu_q^*(\mathbb{S}^q)$. The spaces $X^p(\mathbb{S}^q)$ and $C(\mathbb{S}^q)$ on the sphere are defined analogously to the case of the interval.

A spherical polynomial of degree m is the restriction to $\mathbb{S}^q$ of a polynomial in $q+1$ real variables with total degree m. For integer $n \geq 0$, the class of all spherical polynomials of degree at most n will be denoted by Π_n^q. As before, we extend this notation for non-integer values of n by setting $\Pi_n^q \stackrel{\text{def}}{=} \Pi_{\lfloor n \rfloor}^q$. For integer $\ell \geq 0$, the class of all homogeneous, harmonic, spherical polynomials of degree ℓ will be denoted by $\mathbf{H}_\ell^q$, and its dimension by d_ℓ^q. For each integer $\ell \geq 0$, let $\{Y_{\ell,k} : k = 1, 2, \ldots, d_\ell^q\}$ be a μ_q^*-orthonormalized basis for $\mathbf{H}_\ell^q$. It is known that for any integer $n \geq 0$, the system $\{Y_{\ell,k} : \ell = 0, 1, \ldots, n, \ k = 1, 2, \ldots, d_\ell^q\}$ is an orthonormal basis for Π_n^q, cf. [63,75]. The connection with the theory of orthogonal polynomials on $[-1, 1]$ is the following addition formula

$$\sum_{k=1}^{d_\ell^q} Y_{\ell,k}(\mathbf{x}) Y_{\ell,k}(\mathbf{y}) = \omega_{q-1}^{-1} p_\ell^{(q/2-1,q/2-1)}(1) p_\ell^{(q/2-1,q/2-1)}(\mathbf{x} \cdot \mathbf{y}), \quad \ell = 0, 1, \ldots,$$

cf. [63] where the notation is different.

Analogues of the direct and converse theorems in the case of approximation on the sphere are given, for instance, by Pawelke [71] and Lizorkin and Rustamov [35]. The existence of M–Z quadrature measures was proved in [54]. Numerical constructions and various experiments to demonstrate the effectiveness of the localized operators are given in [34]. A wavelet-like representation including local smoothness classes is given in [47]. In the case of spherical caps, the existence of M–Z quadrature measures is given in [46,9]; the corresponding results on spherical triangles are proved in [44] and numerical constructions are given in [2].

The analogue of the PBF network in this context is the so-called *zonal function* (ZF) *network*. A zonal function network is a function of the form $\mathbf{x} \mapsto \sum_{k=1}^n c_k \phi(\mathbf{x} \cdot \mathbf{y}_k)$, where $\mathbf{x}$ and the $\mathbf{y}_k$'s are on $\mathbb{S}^q$ and $\phi \in L^1(q/2 - 1, q/2 - 1)$. We observe that analogous to the "Mercer expansion"

$$G(\mathbf{x} - \mathbf{y}) = \sum_{\mathbf{k}} \hat{G}(\mathbf{k}) \exp(i\mathbf{k} \cdot \mathbf{x}) \overline{\exp(i\mathbf{k} \cdot \mathbf{y})},$$

for the activation function G of a PBF network, one has the expansion

$$\phi(\mathbf{x} \cdot \mathbf{y}) = \sum_{\ell=0}^{\infty} \hat{\phi}(\ell) \sum_{k=1}^{d_\ell^q} Y_{\ell,k}(\mathbf{x}) Y_{\ell,k}(\mathbf{y})$$

for the activation function ϕ of a ZF network. The ideas in Sect. 4 can be carried over almost verbatim to the case of ZF networks. In particular, the direct and converse theorems in this connection are obtained in [55,56]. The wavelet-like representation for ZF network frames was announced by Shvarts in a joint paper with HNM in a meeting in Barcelona, Spain, in December, 2011.

Acknowledgments We thank the referee for his comments, corrections, critical observations, valuable suggestions, and, in particular, for pointing out [14,32,40,41,72].

References

1. Askey, R., Wainger, S.: A convolution structure for Jacobi series. Am. J. Math. **91**(2), 463–485 (1969)
2. Beckman, J., Mhaskar, H.N., Prestin, J.: Quadrature formulas for integration of multivariate trigonometric polynomials on spherical triangles. Int. J. Geomath. (accepted)
3. Blackmore, K.L., Williamson, R.C., Mareels, I.M.Y.: Decision region approximation by polynomials or neural networks. IEEE Trans. Inf. Theory **43**(3), 903–907 (1997)
4. Carleson, L.: On convergence and growth of partial sums of Fourier series. Acta Math. **116**, 135–157 (1966)
5. Chandrasekharan, S., Jayaraman, K., Mhaskar, H.N.: Minimum Sobolev norm interpolation with trigonometric polynomials on the torus. J. Comput. Phys. (accepted)
6. Cheney, E.W.: Introduction to Approximation Theory, 2nd edn. Chelsea Publishing Company, New York (1982)
7. Chui, C.K., Mhaskar, H.N.: Smooth function extension based on high dimensional unstructured data. Math. Comput. (accepted)
8. Czipszer, J., Freud, G.: Sur l'approximation d'uns fonction periodique et ses derivees successives par un polynomie trigonometrique et par ses derivees successives. Acta Math. **5**, 285–290 (1957)
9. Dai, F., Wang, H.: Positive cubature formulas and Marcinkiewicz–Zygmund inequalities on spherical caps. Constr. Approx. **31**(1), 1–36 (2010)
10. Daubechies, I.: Ten Lectures on Wavelets, CBMS-NSF Series in Appl. Math., SIAM Publ., Philadelphia (1992)
11. DeVore, R., Howard, R., Micchelli, C.A.: Optimal nonlinear approximation. Manuscr. Math. **63**, 469–478 (1989)
12. DeVore, R.A., Lorentz, G.G.: Constructive Approximation. Springer, Berlin (1993)
13. Dick, J., Pillichshammer, F.: Digital Nets and Sequences. Cambridge University Press, Cambridge (2010)
14. Dung, D., Micchelli, C.A.: Sparse grid approximation by translates of the Korobov function (arXiv:1212.6160 [math.FA])

15. Dyn, N.: Interpolation and approximation by radial and related functions. In: Chui, K., Schumaker, L.L., Ward, J.D. (eds.) Approximation Theory VI, vol. 1, pp. 211–234. Academic Press, Boston (1989)

16. Erb, W.: Optimally space localized polynomials with applications in signal processing. J. Fourier Anal. Appl. **18**(1), 45–66 (2012)

17. Erdélyi, A., Magnus, W., Oberhettinger, F., Tricomi, F.G.: Higher Transcendental Functions, vol. 1. McGraw Hill, New York (1953)

18. Fejér, L.: Untersuchungen über Fouriersche Reinen. Math. Ann. **58**, 51–69 (1904)

19. Filbir, F., Mhaskar, H.N., Prestin, J.: On a filter for exponentially localized kernels based on Jacobi polynomials. J. Approx. Theory **160**, 256–280 (2009)

20. Filbir, F., Mhaskar, H.N.: A quadrature formula for diffusion polynomials corresponding to a generalized heat kernel. J. Fourier Anal. Appl. **16**, 629–657 (2010)

21. Filbir, F., Mhaskar, H.N.: Marcinkiewicz–Zygmund measures on manifolds. J. Complex. **27**, 568–596 (2011)

22. Filbir, F., Mhaskar, H.N., Prestin, J.: On the problem of parameter estimation in exponential sums. Constr. Approx. **35**(3), 323–343 (2012)

23. Fröhlich, J., Uhlmann, M.: Orthonormal polynomial wavelets on the interval and applications to the analysis of turbulent flow fields. Siam J. Appl. Math. **63**, 1789–1830 (2003)

24. Girgensohn, R., Prestin, J.: Lebesgue constants for an orthogonal polynomial Schauder basis. J. Comput. Anal. Appl. **2**, 159–175 (2000)

25. Girosi, F., Jones, M., Poggio, T.: Regularization theory and neural networks architectures. Neural Comput. **7**, 219–269 (1995)

26. Hardy, G.H.: Divergent Series, AMS Chelsea Publ., American Mathematical Society, Providence (1991)

27. Hardy, R.L.: Theory and applications of the multiquadric-biharmonic method, 20 years of discovery 1968–1988. Comput. Math. Appl. **19**(8–9), 163–208 (1990)

28. Hassanieh, H., Indyk, P., Katabi, D., Price, E.: Simple and Practical Algorithm for Sparse Fourier Transform, ACM-SIAM Symposium on discrete algorithms, SIAM publ., pp. 1183–1194 (2012)

29. Horn, R.A., Johnson, C.R.: Matrix Analysis. Cambridge University Press, Cambridge (1985)

30. Hunt, R.: On the convergence of Fourier series. Proc. Conf. Edwardsville, pp. 235–255 (1967)

31. Kilgore, T.J., Prestin, J., Selig, K.: Orthogonal algebraic polynomial Schauder bases of optimal degree. J. Fourier Anal. Appl. **2**, 597–610 (1996)

32. Kolmogorov, A.N.: Über die beste Annäherung von Funktionen einer Funktionklasse. Ann. Math. **37**, 107–111 (1936)

33. Lavretsky, E., Hovakimyan, N., Calise, A.: Adaptive Extremum Seeking Control Design. American Control Conference (2003)

34. Le Gia, Q.T., Mhaskar, H.N.: Localized linear polynomial operators and quadrature formulas on the sphere. SIAM J. Numer. Anal. **47**(1), 440–466 (2008)

35. Lizorkin, P.I., Rustamov, Kh.P.: Nikolskii–Besov spaces on the sphere in connection with approximation theory. Tr. Mat. Inst. Steklova **204**, 172–201 (1993) (Proc. Steklov Inst. Math. **3**, 149–172 (1994))

36. Lorentz, G.G., Golitschek, M.v., Makovoz, Y.: Constructive Approximation, Advanced Problems. Springer, Berlin (1996)

37. Lorentz, R.A., Sahakian, A.A.: Orthogonal trigonometric Schauder bases of optimal degree for $C(0, 2\pi)$. J. Fourier Anal. Appl. **1**, 103–112 (1994)

38. Lubinsky, D.S.: Marcinkiewicz inequalities: methods and results. In: Milovanovic, G.V. (ed.) Recent Progress in Inequalities, pp. 213–240. Kluwer, Dordrecht (1998)

39. Maggioni, M., Mhaskar, H.N.: Diffusion polynomial frames on metric measure spaces. Appl. Comput. Harm. Anal. **24**(3), 329–353 (2008)

40. Maiorov, V.: Almost optimal estimates for best approximation by translates on a torus. Constr. Approx. **21**, 1–20 (2005)

41. Mathé, P.: s-Number in information-based complexity. J. Complex. **6**, 41–66 (1990)

42. Mhaskar, H.N.: When is approximation by Gaussian networks necessarily a linear process? Neural Netw. **17**, 989–1001 (2004)

43. Mhaskar, H.N.: Approximation theory and neural networks. In: Jain, K., Krishnan, M., Mhaskar, H.N., Prestin, J., Singh, D. (eds.) Wavelet Analysis and Applications, Proceedings of the international workshop in Delhi, 1999, pp. 247–289. Narosa Publishing, New Delhi (2001)

44. Mhaskar, H.N.: Local quadrature formulas on the sphere, II. In: Neamtu, M., Saff, E.B. (eds.) Advances in Constructive Approximation, pp. 333–344. Nashboro Press, Nashville (2004)

45. Mhaskar, H.N.: Polynomial operators and local smoothness classes on the unit interval. J. Approx. Theory **131**, 243–267 (2004)

46. Mhaskar, H.N.: Local quadrature formulas on the sphere. J. Complex. **20**, 753–772 (2004)

47. Mhaskar, H.N.: On the representation of smooth functions on the sphere using finitely many bits. Appl. Comput. Harmon. Anal. **18**(3), 215–233 (2005)

48. Mhaskar, H.N.: Polynomial operators and local smoothness classes on the unit interval, II. Jaén J. Approx. **1**(1), 1–25 (2005)

49. Mhaskar, H.N.: Eignets for function approximation. Appl. Comput. Harmon. Anal. **29**, 63–87 (2010)

50. Mhaskar, H.N., Micchelli, C.A.: Approximation by superposition of a sigmoidal function and radial basis functions. Adv. Appl. Math. **13**, 350–373 (1992)

51. Mhaskar, H.N., Micchelli, C.A.: Degree of approximation by neural and translation networks with a single hidden layer. Adv. Appl. Math. **16**, 151–183 (1995)

52. Mhaskar, H.N., Micchelli, C.A.: Dimension-independent bounds on approximation by neural networks. IBM J. Res. Dev. **38**, 277–284 (1994)

53. Mhaskar, H.N., Micchelli, C.A.: How to choose an activation function. In: Cowan, J.D., Tesauro, G., Alspector, J. (eds.) Neural Information Processing Systems, vol. 6, pp. 319–326. Morgan Kaufmann Publishers, San Fransisco (1993)

54. Mhaskar, H.N., Narcowich, F.J., Ward, J.D.: Spherical Marcinkiewicz–Zygmund inequalities and positive quadrature. Math. Comput. **70**(235), 1113–1130 (2001) (Corrigendum. Math. Comput. **71**, 453–454 (2001))

55. Mhaskar, H.N., Narcowich, F.J., Ward, J.D.: Approximation properties of zonal function networks using scattered data on the sphere. Adv. Comput. Math. **11**, 121–137 (1999)

56. Mhaskar, H.N., Narcowich, F.J., Prestin, J., Ward, J.D.: L^p Bernstein estimates and approximation by spherical basis functions. Math. Comput. **79**(271), 1647–1679 (2010)

57. Mhaskar, H.N., Pai, D.V.: Fundamentals of Approximation Theory. Alpha Science Intl., Revised Edition (2007)

58. Mhaskar, H.N., Prestin, J.: On the detection of singularities of a periodic function. Adv. Comput. Math. **12**, 95–131 (2000)

59. Mhaskar, H.N., Prestin, J.: Polynomial frames: a fast tour. In: Chui, K., Neamtu, M., Schumaker, L. (eds.) Approximation Theory XI: Gatlinburg 2004, Mod. Methods Math, pp. 287–318. Nashboro Press, Brentwood (2005)

60. Mhaskar, H.N., Prestin, J.: On local smoothness classes of periodic functions. J. Fourier Anal. Appl. **11**(3), 353–373 (2005)

61. Mhaskar, H.N., Prestin, J.: Polynomial operators for spectral approximation of piecewise analytic functions. Appl. Comput. Harmon. Anal. **26**, 121–142 (2009)

62. Micchelli, C.A.: Interpolation of scattered data: distance matrices and conditionally positive definite functions. Constr. Approx. **2**(1), 11–22 (1986)

63. Müller, C.: Spherical Harmonics, Lecture Notes in Mathematics, vol. 17. Springer, Berlin (1966)

64. Narcowich, F.J., Ward, J.D.: Norms of inverses and condition numbers for matrices associated with scattered data. J. Approx. Theory **64**, 69–94 (1991)

65. Narcowich, F.J., Ward, J.D.: Norm estimates for inverses of a general class of scattered data radial function interpolation matrices. J. Approx. Theory **69**, 84–109 (1992)

66. Natanson, I.P.: Constructive Function Theory, vol. I. Frederick Ungar, New York (1964)

67. Nevai, P.: Mean convergence of Lagrange interpolation. J. Approx. Theory **18**, 363–377 (1976)

68. Nevai, P.: Orthogonal Polynomials. Mem. Amer. Math. Soc., vol. 203. American Mathematical Society, Providence (1979)

69. Nevai, P.: The incredible but, I swear, true story of n vs. $2n$ in the Bernstein Inequality (2013, in preparation)

70. Park, J., Sandberg, I.W.: Universal approximation using radial basis function networks. Neural Comput. **3**, 246–257 (1991)

71. Pawelke, S.: Über die Approximationsordnung bei Kugelfunktionen und algebraischen Polynomen. Tôhoku Math. J. **24**, 473–486 (1972)

72. Pinkus, A.: n-Widths in Approximation Theory. Springer, Berlin (1985)

73. Schaback, R., Wendland, H.: Inverse and saturation theorems for radial basis function interpolation. Math. Comput. **71**, 669–681 (2002)

74. Stechkin, S.B.: The approximation of periodic functions by Fejr sums. (Russian). Trudy Mat. Inst. Steklov. **62**, 48–60 (1961)
75. Stein, E.M., Weiss, G.: Fourier Analysis on Euclidean Spaces. Princeton University Press, Princeton (1971)
76. Szegő, G.: Orthogonal Polynomials, Amer. Math. Soc. Colloq. Publ., vol. 23. American Mathematical Society, Providence (1975)
77. Timan, A.F.: Theory of Approximation of Functions of a Real Variable. Pergamon Press, Oxford (1963; English translation)
78. Zeevi, A., Meir, R., Maiorov, V.: Approximation and estimation bounds for nonlinear regression using mixtures of experts. Neural Netw. **10**(1), 99–109 (1997)
79. Zygmund, A.: Trigonometric Series. Cambridge University Press, Cambridge (1977)

6

On n-layered $QTAG$-modules

Fahad Sikander · Ayazul Hasan · Alveera Mehdi

Abstract A module M over an associative ring with unity is a $QTAG$-module if every finitely generated submodule of any homomorphic image of M is a direct sum of uniserial modules. There are many fascinating concepts related to these modules. Here we introduce the notion of n-layered $QTAG$-modules and discuss some interesting properties of these modules. We show that a $QTAG$-module M is n-layered if and only if M/N is an n-layered module, whenever N is a finitely generated submodule of M and $n \geq 1$ is an integer.

Keywords QTAG-module · ω-elongation · Totally projective modules · $(\omega + k)$-projective modules

Mathematics Subject Classification (2010) 16K20

Communicated by S.K. Jain.

F. Sikander (✉)
College of Computing and Informatics, Saudi Electronic University (Jeddah Branch),
Jeddah 23442, Kingdom of Saudi Arabia
e-mail: fahadsikander@gmail.com; f.sikander@seu.edu.sa

A. Hasan
Department of Mathematics, Integral University, Lucknow 226026, India
e-mail: ayaz.maths@gmail.com

A. Mehdi
Department of Mathematics, Aligarh Muslim University, Aligarh 202002, India
e-mail: alveera_mehdi@rediffmail.com

1 Introduction and preliminaries

The study of QTAG-modules was initiated by Singh [8]. Mehdi et al. [4] worked a lot on these modules. They studied different notions and structures of QTAG-modules and developed the theory of these modules by introducing several notions, investigated some interesting properties and characterized them. Yet there is much to explore.

Throughout this paper, all rings are associative with unity and modules M are unital $QTAG$-modules. An element $x \in M$ is uniform, if xR is a non-zero uniform (hence uniserial) module and for any R-module M with a unique composition series, $d(M)$ denotes its composition length. For a uniform element $x \in M$, $e(x) = d(xR)$ and $H_M(x) = \sup\{d(\frac{yR}{xR}) \mid y \in M, \ x \in yR$ and y uniform$\}$ are the exponent and height of x in M, respectively. $H_k(M)$ denotes the submodule of M generated by the elements of height at least k and $H^k(M)$ is the submodule of M generated by the elements of exponents at most k. A submodule N of M is h-pure in M if $N \cap H_k(M) = H_k(N)$, for every integer $k \geq 0$. A submodule N of a QTAG-module M is height finite, if the heights of the elements of N take only finitely many values. M is h-divisible if $M = M^1 = \bigcap_{k=0}^{\infty} H_k(M)$ and it is h-reduced if it does not contain any h-divisible submodule. In other words it is free from the elements of infinite height.

A submodule $N \subset M$ is nice [3, Definition 2.3] in M, if $H_\sigma(M/N) = (H_\sigma(M) + N)/N$ for all ordinals σ, i.e. every coset of M modulo N may be represented by an element of the same height.

A family of nice submodules $\mathcal{N}$ of submodules of M is called a nice system in M if

(i) $0 \in \mathcal{N}$;
(ii) If $\{N_i\}_{i \in I}$ is any subset of $\mathcal{N}$, then $\Sigma_I N_i \in \mathcal{N}$;
(iii) Given any $N \in \mathcal{N}$ and any countable subset X of M, there exists $K \in \mathcal{N}$ containing $N \cup X$, such that K/N is countably generated [4].

A h-reduced $QTAG$-module M is called totally projective if it has a nice system.

For a $QTAG$-module M, there is a chain of submodules $M^0 \supset M^1 \supset M^2 \cdots \supset M^\tau = 0$, for some ordinal τ. $M^{\sigma+1} = (M^\sigma)^1$, where M^σ is the σth-Ulm submodule of M. A fully invariant submodule $L \subset M$ is a large submodule of M, if $L + B = M$ for every basic submodule B of M. Several results which hold for TAG-modules also hold good for $QTAG$-modules [8]. Notations and terminology are follows from [1,2].

2 n-Layered $QTAG$-modules and its properties

Recall that a $QTAG$-module M is $(\omega + 1)$-projective if there exists submodule $N \subset H^1(M)$ such that M/N is a direct sum of uniserial modules and a $QTAG$ module M is $(\omega + k)$-projective if there exists a submodule $N \subset H^k(M)$ such that M/N is a direct sum of uniserial modules [4].

Let σ be a limit ordinal such that $\sigma = \omega + \beta$. A $QTAG$-module M is called σ-projective, if there exists a submodule $N \subset H^\beta(M)$ such that M/N is a direct sum of uniserial modules. A QTAG-module M is totally projective, if and only if $M/H_\sigma(M)$ is σ-projective for every ordinal σ.

A $QTAG$-module is an ω-elongation of a totally projective $QTAG$-module by a $(\omega + k)$-projective $QTAG$-module if and only if $H_\omega(M)$ is totally projective and $M/H_\omega(M)$ is $(\omega + k)$-projective.

A $QTAG$-module M is a *strong ω-elongation* of a totally projective module by a $(\omega+n)$-projective module if $H_\omega(M)$ is totally projective and there exists $N \subseteq H^n(M)$ such that $\frac{M}{N+H_\omega(M)}$ is a direct sum of uniserial modules [5].

Referring to our criterion from [7], M is a Σ-module or layered module if $\text{Soc}(M) = \bigcup_{k<\omega} M_k$, where $M_k \subseteq M_{k+1} \subseteq \text{Soc}(M)$ and for every k, $M_k \cap H_k(M) = \text{Soc}(H_\omega(M))$.

In [5], it was shown that any $(\omega + 1)$-projective σ-module is a direct sum of countable modules of length almost $(\omega + 1)$. Moreover, we extended this assertion to the so called strong ω-elongations. It was established that any strong ω-elongation of a totally projective module by a $(\omega + 1)$-projective module is a Σ-module precisely when it is totally projective.

That is why it naturally comes under what additional conditions on the module structure this type of results hold for every $n \in \mathbb{N}$. To achieve this goal we state the following new concept, which is a generalization of the corresponding one for Σ-module.

Definition 1 A $QTAG$-module M is said to be *n-layered module* if for some $n < \omega$, $H^n(M) = \bigcup_{k<\omega} M_k$, $M_k \subseteq M_{k+1} \subseteq H^n(M)$ and for all $k \geq 1$, $M_k \cap H_k(M) = H^n(H_\omega(M))$.

Remark 1 Equivalently, we may say that M is a n-layered module if and only if $H^n(M) = \bigcup N_k$, $N_k \subseteq N_{k+1} \subseteq H^n(M)$ and for every $k \geq 1$, $N_k \cap H_k(M) \subseteq H_\omega(M)$.

Also, $N_k \subseteq N_k + H^n(H_\omega(M))$ implies that $H^n(M) = \bigcup(N_k + H^n(H_\omega(M)))$ and $(N_k + H^n(H_\omega(M))) \cap H_k(M) = H^n(H_\omega(M)) + (N_k \cap H_k(M)) = H^n(H_\omega(M))$. Therefore $M_k = N_k + H^n(H\omega(M))$ and $N_k \cap H_k(M) \subseteq H_\omega(M)$, equivalently $N_k \cap H_k(M) = H^n(H_\omega(M))$.

Remark 2 Every layered module is 1-layered module and vice-versa. Since $H^n(M) \subseteq H^m(M)$, for $n \leq m$, every m-layered module is a n-layered module.

Now we investigate some properties of n-layered modules.

Lemma 1 *For $n \geq 1$, h-pure submodules of n-layered modules are n-layered modules. Moreover, the submodules of n-layered modules with the same first Ulm submodules are n-layered.*

Proof Let M be a n-layered $QTAG$-module such that $H^n(M) = \bigcup_{j<\omega} M_j$, $M_j \subseteq M_{j+1} \subseteq H^n(M)$ and $M_j \cap H_j(M) \subseteq H_\omega(M)$. Now for any h-pure submodule N of M, $H^n(N) = \bigcup_{j<\omega} N_j$, where $N_j = M_j \cap N$ and

$$N_j \cap H_j(N) \subseteq N \cap H_\omega(M) = H_\omega(N)$$

and the result follows.

If K is an arbitrary submodule of M such that $H_\omega(K) = H_\omega(M)$, then $H^n(K) = \bigcup_{j<\omega} K_j$, where $K_j = M_j \cap K$ and

$$K_j \cap H_j(K) \subseteq K \cap H_\omega(M) = H_\omega(K)$$

and we are done.

Lemma 2 *Let N be submodule of a h-reduced module M and $n \geq 1$. Then M is n-layered if and only if N is n-layered, where N is a $H_{\omega+n-1}(M)$-high submodule of M.*

Proof For any ordinal α, $H_\alpha(M)$-high submodules of M are h-pure in M. Now if N is a $H_{\omega+n-1}(M)$-high submodule of M, it is h-pure in M and by Lemma 1, if M is n-layered, then N is also n-layered.

We have "Let M be a h-reduced QTAG-module and let N be a $H_{\alpha+k}$-high submodule of M with α a limit ordinal and $k \geq 1$. Then $H^n(M) = H^n(N) \oplus H^n(K)$ for $n > k$ and any complementary summand K of a maximal summand of $H_\alpha(M)$ bounded by k" [4].

Now for the converse, we have $H^n(M) = H^n(N) \oplus H^n(K)$, where K is a $H_{n-1}(H_\omega(M))$-high submodule of $H_\omega(M)$. Since $H^n(N) = \bigcup_{j<\omega} N_j$, $N_j \subseteq N_{j+1} \subseteq H^n(N)$ and $N_j \cap H_j(N) \subseteq H_\omega(N)$, by defining $M_j = N_j \oplus H^n(K)$, we have $H^n(M) = \bigcup_{j<\omega} M_j$. Since N is h-pure in M and $K \subseteq H_\omega(M)$, $M_j \cap H_j(M) = H^n(K) + (N_j \cap H_j(M)) = H^n(K) + (N_j \cap H_j(N)) \subseteq H_\omega(M)$.

Proposition 1 *For $n \geq 1$, all Σ-modules with h-divisible first Ulm submodule are n-layered modules.*

Proof Let M be a Σ-module such that $H_\omega(M)$ is h-divisible. Since h-divisible submodules are direct summands, we have $M = H_\omega(M) \oplus N$, where N is contained in a high submodule of M, hence N is a direct sum of uniserial submodules. Again $\frac{M}{H_\omega(M)} \simeq N$ and we are done.

Proposition 2 *Direct sums of n-layered modules are n-layered modules.*

Proof Let M be a direct sum of n-layered modules such that $M = \bigoplus_{j\in J} N_j$. Here N_j's are n-layered modules. Therefore $H^n(N_j) = \bigcup_{i<\omega} N_{ij}$, $N_{ij} \subseteq N_{(i+1)j} \subseteq N_j$ and $N_{ij} \cap H_i(N_j) \subseteq H_\omega(N_j)$ for $i < \omega$, $j \in J$.

Furthermore, $H^n(M) = \bigoplus_{j\in J} H^n(N_j) = \bigoplus_{j\in J} (\bigcup_{i<\omega} N_{ij}) = \bigcup_{i<\omega} (\bigoplus_{j\in J} N_{ij}) = \bigcup_{i<\omega} M_i$, where $M_i = \bigoplus_{j\in J} N_{ij}$ and

$$
\begin{aligned}
M_i \cap H_i(M) &= \left(\bigoplus_{j\in I} N_{ij}\right) \cap \left(\bigoplus_{j\in I} H_i(N_j)\right) \\
&= \bigoplus_{j\in I} (N_{ij} \cap H_i(N_j)) \\
&\subseteq \bigoplus_{j\in I} H_\omega(N_j)
\end{aligned}
$$

$$= H_\omega \left(\bigoplus_{j \in I} N_j \right)$$

$$= H_\omega(M)$$

and the result follows.

Proposition 3 *For $k \geq 1$, M is a n-layered module if and only if $H_k(M)$ is n-layered.*

Proof If M is a n-layered module, then $H^n(M) = \bigcup_{j < \omega} N_j$, $N_j \subseteq N_{j+1} \subseteq H^n(M)$ and $N_j \cap H_j(M) \subseteq H_\omega(M)$, for every $j < \omega$. Therefore $H^n(H_k(M)) = \bigcup_{i < \omega} T_j$, where $T_j = N_j \cap H_k(M)$ and

$$
\begin{aligned}
T_j \cap H_{k+j}(M) &= T_j \cap H_j(H_k(M)) \\
&\subseteq N_j \cap H_j(M) \\
&\subseteq H_\omega(M) \\
&= H_\omega(H_k(M)).
\end{aligned}
$$

Thus $H_k(M)$ is also n-layered.

For the converse, suppose $H_k(M)$ is n-layered. If $k=1$, then $H_1(M)$ is n-layered and we have $H^n(H_1(M)) = \bigcup_{j < \omega} T_j$, $T_j \subseteq T_{j+1} \subseteq H^n(H_1(M))$ and $T_j \cap H_{j+1}(M) \subseteq H_\omega(M)$. Let $S = \{x \mid x \in H^n(M), x \notin H^n(H_1(M))\}$. Then $H^n(M) = S \cup H^n(H_1(M))$. Define $K_j \subseteq M$ such that $K_j \cap H_1(M) = \phi$ and $H^n(M \backslash H_1(M)) = \bigcup_{j < \omega} K_j$ such that $\langle K_j \rangle \cap H_1(M) \subseteq T_j$. This implies that $H^n(M) = \bigcup (T_j + \langle K_j \rangle)$, and

$$
\begin{aligned}
(T_j + \langle K_j \rangle) \cap H_{j+1}(M) &\subseteq (T_j + \langle K_j \rangle) \cap H_1(M) \\
&= T_j + (\langle K_j \rangle \cap H_1(M)) \\
&= T_j.
\end{aligned}
$$

Therefore $(T_j + \langle K_j \rangle) \cap H_{j+1}(M) \subseteq T_j \cap H_{j+1}(M) \subseteq H_\omega(M)$ and the result follows.

Proposition 4 *A QTAG-module M is n-layered module if and only if its large submodule L is n-layered.*

Proof For a large submodule L of M, $H_\omega(L) = H_\omega(M)$ [6]. Therefore by Lemma 1, L is n-layered whenever M is n-layered.

Conversely, suppose L is n-layered such that $L = \sum_{k < \omega} H^k(H_{m_k}(M))$, where $m_1 \leq m_2 \leq \cdots \leq m_k$ is a monotonically increasing sequence of positive integers. Now $H^n(L) = Soc(H_{m_1}(M) + \cdots + H^n(H_{m_n}(M)))$ therefore

$$H^n(H_{m_n}(M)) \subseteq H^n(L) \subseteq H^n(H_{m_1}(M)).$$

Also $H^n(L) = \bigcup_{j < \omega} L_j$, $L_j \subseteq L_{j+1} \subseteq H^n(L)$ and $L_j \cap H_j(L) \subseteq H_\omega(L)$ and $H^n(H_{m_n}(M)) = \bigcup_{j < \omega} N_j$, where $N_j = L_j \cap H^n(H_{m_n}(M)) = L_j \cap (H_{m_n}(M))$. Again $N_j \cap H_{t_j}(M) \subseteq N_j \cap H_j(L) \subseteq H_\omega(L) = H_\omega(M)$, for some $t_j \geq \max(j, m_n)$ with $H^n(H_{t_j}(M)) \subseteq H^n(H_j(L))$, as $H_j(L)$ is also a large submodule of M. Now $H_{m_n}(M)$ is n-layered module and by Proposition 3, M is also n-layered module.

Proposition 5 *Let N be a submodule of M such that M/N is bounded. Then M is n-layered module if and only if N is n-layered module.*

Proof Since M/N is bounded, then there exists an integer k such that $H_k(M/N) = 0$ or $H_k(M) \subseteq N$. Therefore $H_\omega(H_k(M)) = H_\omega(M) = H_\omega(N)$ and by Lemma 1, if M is n-layered then N is also n-layered.

Conversely, if N is a n-layered module then by Lemma 1, $H_k(M) \subseteq N$ is also n-layered. Therefore by Proposition 3, M is also n-layered.

Proposition 6 *Let N be a height-finite, submodule of M. If M/N is n-layered, then M is n-layered.*

Proof Since M/N is n-layered, $H^n(M/N) = \bigcup_{j<\omega}(K_j/N) = (\bigcup K_j)/N$, where $K_j \subseteq K_{j+1} \subseteq M$ and $\left(\frac{K_j}{N}\right) \cap H_j\left(\frac{M}{N}\right) \subseteq H_\omega\left(\frac{M}{N}\right)$. Now N is height-finite, therefore nice in M and $K_j \cap H_j(M) \subseteq H_\omega(M) + N$. There exists a positive integer $t_j \geq j$ such that $N \cap H_{t_j}(M) \subseteq H_\omega(M)$. Also

$$K_j \cap H_{t_j}(M) \subseteq (H_\omega(M) + N) \cap H_{t_j}(M)$$
$$= H_\omega(M) + (N \cap H_{t_j}(M))$$
$$= H_\omega(M).$$

Now $\left(\frac{H^n(M)+N}{N}\right) \subseteq H^n(M/N)$ and $H^n(M) \subseteq \bigcup_{j<\omega} K_j$. Thus $H^n(M) = \bigcup_{j<\omega} T_j$, where $T_j = K_j \cap H^n(M) = H^n(K_j)$ and the result follows.

Remark 3 Let N be a height-finite submodule of M. If M/N is a Σ-module, then M is also a Σ-module.

Proposition 7 *Let N be a submodule of M.*

(i) *if $N \cap H_n(M) = H_n(N)$ and N is finitely generated or $N \subseteq H_\omega(M)$ and M is n-layered, then M/N is also n-layered;*

(ii) *if $N \subseteq H^k(M)$, for some $k \geq 1$ and either N is finitely generated or $N \subseteq H_\omega(M)$ and M is $(n+k)$-layered, then M/N is also n-layered.*

Proof (i) If M is n-layered, then $H^n(M) = \bigcup_{j<\omega} M_j$, $M_j \subseteq M_{j+1}$ and $M_j \cap H_j(M) \subseteq H_\omega(M)$. Now, $H^n\left(\frac{M}{N}\right) = \left(\frac{H^n(M)+N}{N}\right) = \bigcup_{j<\omega}\left(\frac{M_j+N}{N}\right)$. Therefore, $\left(\frac{M_j+N}{N}\right) \cap H_j\left(\frac{M}{N}\right) = \frac{[N+((M_j+N)\cap H_j(M))]}{N}$.

When $N \subseteq H_j(M)$, for every positive integer j, then

$$(M_j + N) \cap H_j(M) \subseteq N + (M_j \cap H_j(M)) \subseteq N + H_\omega(N).$$

Since $\left(\frac{N+H_\omega(M)}{N}\right) \subseteq H_\omega\left(\frac{M}{N}\right)$, the result follows.

When N is finitely generated, then there exists an integer $t_j \geq j$ such that $(M_j + N) \cap H_{t_j}(M) \subseteq H_\omega(M)$. Therefore $\left(\frac{M_j+N}{N}\right) \cap H_{t_j}\left(\frac{M}{N}\right) \subseteq \left(\frac{N+H_\omega(M)}{N}\right) = H_\omega\left(\frac{M}{N}\right)$ and we are done.

If $N \subseteq H_\omega(M)$ and we have $H^n \left(\frac{M}{N}\right) = \bigcup_{j<\omega} \left(\frac{M_j}{N}\right)$, $M_j \subseteq M_{j+1} \subseteq M$, where $(M_j/N) \cap H_j(M/N) \subseteq H_\omega(M/N) = H_\omega(M)/N$. Therefore $M_j \cap H_j(M) \subseteq H_\omega(M)$. Since $\left(\frac{H^n(M)+N}{N}\right) \subseteq H^n \left(\frac{M}{N}\right)$, $H^n(M) = \bigcup_{j<\omega} H^n(M_j)$ and the result follows.

(ii) Since $H^n \left(\frac{M}{N}\right) \subseteq \left(\frac{H^{n+k}(M)+N}{N}\right)$, we are through.

Proposition 8 *If for some ordinal α, $M/H_\alpha(M)$ is n-layered, then M is n-layered.*

Proof We have $H^n(M/H_\alpha(M)) = \bigcup_{j<\omega}(M_j/H_\alpha(M))$, $M_j \subseteq M_{j+1} \subseteq M$, $M_j \cap H_j(M) \subseteq H_\omega(M)$ for every $j < \omega$. Now

$$\left(\frac{H^n(M) + H_\alpha(M)}{H_\alpha(M)}\right) \subseteq H^n \left(\frac{M}{H_\alpha(M)}\right)$$

therefore $H^n(M) \subseteq \bigcup_{j<\omega} M_j$. If we put $T_j = H^n(M_j)$, then $H^n(M) = \bigcup_{j<\omega} T_j$. But $T_j \cap H_j(M) \subseteq M_j \cap H_j(M) \subseteq H_\omega(M)$ and we are done.

Now we are in the state to prove our main result which motivated this article.

Theorem 1 *The QTAG-module M is a n-layered module which is a strong ω-elongation of a totally projective module by a $(\omega + n)$-projective module if and only if M is a totally projective module.*

Proof Since M is a strong ω-elongation, $H_\omega(M)$ is totally projective and there exists a submodule $N \subseteq H^n(M)$ such that $\frac{M}{N+H_\omega(M)}$ is a direct sum of uniserial modules and $\frac{M}{N+H_\omega(M)} \simeq \frac{M/H_\omega(M)}{(N+H_\omega(M))/H_\omega(M)}$. Now by the definition of n-layered modules, $H^n(M) = \bigcup_{j<\omega} M_j$, $M_j \subseteq M_{j+1} \subseteq H^n(M)$ and $M_j \cap H_j(M) = H^n(H_\omega(M))$, for every j. Since $N \subseteq H^n(M)$, $N = \bigcup_{j<\omega} N_j$ where $N_j = N \cap M_j$ and $\left(\frac{N+H_\omega(M)}{H_\omega(M)}\right) = \bigcup_{j<\omega} \left(\frac{(N_j+H_\omega(M))}{H_\omega(M)}\right)$.

Now,

$$\left(\frac{(N_j + H_\omega(M))}{H_\omega(M)}\right) \cap H_j \left(\frac{M}{H_\omega(M)}\right) = \frac{(N_j + H_\omega(M))}{H_\omega(M)} \cap \left(\frac{H_j(M)}{H_\omega(M)}\right),$$

$$= \frac{[(N_j + H_\omega(M)) \cap H_j(M)]}{H_\omega(M)},$$

$$= \frac{(H_\omega(M) + (N_j \cap H_j(M)))}{H_\omega(M)},$$

$$= 0.$$

Therefore $M/H_\omega(M)$ is a direct sum of uniserial modules and M is totally projective.

We have shown that if N is a finite submodule of M such that $N \cap H_n(M) = H_n(N)$, then M is an n-layered module if and only if M/N is an n-layered module. Moreover, in [7] we showed that M is Σ-module if and only if M/N is a Σ-module.

We generalize this assertion to n-layered modules for an arbitrary natural number n. For doing this, we need following technical lemmas:

Lemma 3 *Let N be a finitely generated submodule of M. Then for an integer $n \geq 1$,*

$$H^n(M/N) = \frac{H^n(M) + K}{N}$$

where K is a finitely generated submodule of M with $H_n(K) \subseteq N \subseteq K$.

Proof Let $x + N \in H^k(M/N)$ for some $x \in M$ such that there exists $y \in N$ with $d\left(\frac{xR}{yR}\right) = n$. We may express $N \cap H_n(M) = \sum_{i=1}^{m} x_i R$ for some $m \in Z^+$ and put $K = N + \sum y_i R$, where $d\left(\frac{y_i R}{x_i R}\right) = n$. If $y_k \in M$ such that $k \neq 1, \ldots, m$ and $x_k \in N$ such that $d\left(\frac{y_k R}{x_k R}\right) = n$, then $x_k R = x_i R$ for some $i \in \{1, 2, \ldots, m\}$. Therefore $y_k R \subseteq y_i R + H^n(M) \subseteq K + H^n(M)$. The converse is trivial and the result follows.

Lemma 4 *Let K be a h-finite submodule of M having only finite heights in M. If N is a finitely generated submodule of M then $N + K$ is also h-finite assuming finite heights only.*

Proof Since the elements of K assumes only finite number of finite heights, $K \cap H_k(M) \subseteq M^1$, for some $k \geq 1$. Now N is finitely generated submodule and we may express N as $\sum_{i=1}^{m} x_i R$. Consider the submodule N' of N where $N' = \sum_{i=1}^{t} x_i R$ such that $x_i + y_i \in H_{n_i}(M)$ but $x_i + y_i \notin H_{n_i+1}(M)$ with $n_i > k$, $\forall i = 1, 2, \ldots, t$ for some $y_i \in K$. Therefore for each $y \in K$ we have $y + x_i = y - y_i + y_i + x_i \notin H_{n_i+1}(M)$, otherwise $y - y_i + y_i + x_i \in H_{n_i}(M)$ implying that $y - y_i \in H_k(M)$ and $y - y_i \in M^1$. Therefore $y_i + x_i \in H_{n_i+1}(M)$ which is a contradiction whenever $1 \leq i \leq t$. If we put $n = \max\{n_1 + 1, \ldots, n_t + 1\}$, $y + x_i \notin H_n(M)$. Since $y + x_j \notin H_n(M)$ for $t + 1 \leq j \leq n$, we are done.

Now we are ready to prove our main result:

Theorem 2 *For each natural number n, a QT AG-module M is n-layered if and only if M/N is n-layered, where N is a finitely generated submodule of M.*

Proof Suppose that M is an n-layered QTAG-module, then $H^n(M) = \bigcup_{i < \omega} M_i$, $M_i \subseteq M_{i+1} \subseteq H^n(M)$ and, for all $i < \omega$ $M_i \cap H_i(M) \subseteq M^1$. By Lemma 3 we may write $H^n(M/N) = (H^n(M) + K)/N$, for some finitely generated submodule K of M containing N. Furthermore, $H^n(M/N) = \bigcup_{i < \omega} ((M_i + K)/N)$ and by Lemma 4, we calculate that

$$\left(\frac{M_i + K}{N}\right) \cap H_{t_i}\left(\frac{M}{N}\right) = \left(\frac{M_i + K}{N}\right) \cap \left(\frac{H_{t_i}(M) + N}{N}\right)$$

$$= \frac{(M_i + K) \cap (H_{t_i}(M) + N)}{N}$$

$$= \frac{(M_i + K) \cap H_{t_i}(M) + N}{N} \subseteq \frac{M^1 + N}{N} \subseteq \left(\frac{M}{N}\right)^1$$

for every i and some natural number $t_i \geq i$, implying that M/N is an n-layered module.

For reverse implication, suppose that M/N is an n-layered module. Now write $H^n(M/N) = \bigcup_{i<\omega}\left(\frac{T_i}{N}\right)$, $T_i \subseteq T_{i+1} \subseteq M$ and for all $i < \omega$,

$$\left(\frac{T_i}{N}\right) \cap H_i\left(\frac{M}{N}\right) = \left(\frac{M}{N}\right)^1.$$

Since N is finitely generated it is nice in M. Now we may say $\frac{H^n(M)+N}{N} \subseteq H^n\left(\frac{M}{N}\right)$, $\bigcup_{i<\omega}\left(\frac{T_i}{N}\right) = \frac{\bigcup_{i<\omega} T_i}{N}$ and $\left(\frac{M}{N}\right)^1 = \frac{M^1 + N}{N}$. Therefore $H^n(M) = \bigcup_{i<\omega} H^n(T_i)$ and $\left(\frac{T_i}{N}\right) \cap \left(\frac{H_i(M)+N}{N}\right) = \frac{M^1+N}{N}$.
Therefore

$$\frac{(T_i \cap H_i(M) + N)}{N} = \frac{M^1 + N}{N} \quad \text{and}$$
$$\frac{N + (T_i \cap H_i(M))}{N} = \frac{M^1 + N}{N}$$
$$\Rightarrow T_i \cap H_i(M) \subseteq M^1 + N.$$

Since N is finitely generated so there exists $m \in \mathbb{N}$ such that $N \cap H_m(M) \subseteq M^1$, therefore

$$T_i \cap H_{t_i}(M) \subseteq (M^1 + N) \cap H_m(M)$$
$$\subseteq M^1 + N \cap H_m(M) = M^1,$$

for every i and $t_i = m + i$, implying that M is also n-layered.

Acknowledgments   The authors are thankful to the referee for his/her valuable suggestions.

References

1. Fuchs, L.: Infinite Abelian Groups, vol. I. Academic Press, New York (1970)
2. Fuchs, L.: Infinite Abelian Groups, vol. II. Academic Press, New York (1973)
3. Mehdi, A., Abbasi, M.Y., Mehdi, F.: Nice decomposition series and rich modules. South East Asian J. Math. Math. Sci. **4**(1), 1–6 (2005)
4. Mehdi, A., Abbasi, M.Y., Mehdi, F.: On (ω + n)-projective modules. Ganita Sandesh **20**(1), 27–32 (2006)
5. Mehdi, A., Skander, F., Naji Sabah, A.R.K.: On elongations of QTAG-modules. Math. Sci. (accepted for publication).
6. Mehran Hefzi, A., Singh, S.: Ulm-Kaplansky invariants of TAG-modules. Commun. Algebra **13**(2), 355–373 (1985)
7. Naji, S.A.R.K.: A study of different structures of $QTAG$-modules. Ph.D. Thesis, A.M.U., Aligarh (2011)
8. Singh, S.: Some decomposition theorems in abelian groups and their generalizations. In: Ring Theory: Proceedings of Ohio University Conference, vol. 25, pp. 183–189. Marcel Dekker, NY (1976)

Permissions

List of Contributors

Rostislav Grigorchuk
Department of Mathematics, Mailstop 3368, Texas A&M University, College Station, TX 77843-3368, USA

Dhaval Thakkar · Ruchi Das
Department of Mathematics, Faculty of Science, The M. S. University of Baroda, Vadodara 390002, India

Zbigniew Błocki
Instytut Matematyki, Uniwersytet Jagiello´nski, Łojasiewicza 6, 30-348 Krakow, Poland

P. L. Butzer and R. L. Stens
Lehrstuhl A für Mathematik, RWTH Aachen University, 52056 Aachen, Germany

M. M. Dodson
Department of Mathematics, University of York, York YO1O 5DD, UK

P. J. S. G. Ferreira
IEETA/DETI, Universidade de Aveiro, 3810-193 Aveiro, Portugal

J. R. Higgins
I.H.P., 4 rue du Bary, 11250 Montclar, France

G. Schmeisser
Department of Mathematics, University of Erlangen-Nuremberg, 91058 Erlangen, Germany

Hrushikesh N. Mhaskar
Department of Mathematics, California Institute of Technology, Pasadena, CA 91125, USA
Institute of Mathematical Sciences, Claremont Graduate University, Claremont, CA 91711, USA

Paul Nevai
King Abdulaziz University, Jeddah, Saudi Arabia

Eugene Shvarts
Shvarts Scientific Services, Pacoima, CA 91331, USA
Department of Mathematics, University of California at Davis, Davis, CA 95616, USA

Fahad Sikander
College of Computing and Informatics, Saudi Electronic University (Jeddah Branch), Jeddah 23442, Kingdom of Saudi Arabia

Ayazul Hasan
Department of Mathematics, Integral University, Lucknow 226026, India

Alveera Mehdi
Department of Mathematics, Aligarh Muslim University, Aligarh 202002, India

www.ingramcontent.com/pod-product-compliance
Lightning Source LLC
Chambersburg PA
CBHW050449200326

41458CB00014B/5115